Cumulative Impact and Other Disruption Claims in Construction

Richard J. Long, P.E.
Rod C. Carter, CCP, PSP
Harold E. Buddemeyer

LONG INTERNATIONAL

Published in 2014 by Virtualbookworm.com Publishing Inc., College Station, Texas 77842, United States of America.

ISBN 978-1-62137-527-2 (casebound); 978-1-62137-739-9 (electronic).

Library of Congress Number on file with Publisher.

1 2 3 4 5 6 7 8 9 10

Table of Contents

Table of Contents
(continued)

Table of Contents
(continued)

Table of Contents
(continued)

Table of Contents
(continued)

Table of Contents
(continued)

List of Tables

Table of Contents
(continued)

List of Figures

Table of Contents
(continued)

Preface

Construction contracts usually contain a "Changes" clause by which the owner can bilaterally or unilaterally request changes to the scope of work that is to be performed by the contractor, and these requests are typically converted into change orders. However, the owner's actions or inactions can also result in constructive changes to the contract. Change orders to adjust the contract price and time for completion result from a wide range of owner responsible events, including but not limited to, owner-directed increases or decreases in the scope of work to be performed by the contractor, owner-directed changes in the means and methods of the contractor's performance or the materials or equipment to be installed, owner-directed changes in the contractor's planned sequence in which the work is to be performed, design changes, changes in the performance specifications, differing site conditions, constructability issues, late responses to the contractor's properly prepared submittals and requests for information, delays in the delivery of owner-supplied materials and equipment, failure to secure permits in a timely manner, owner interference with the contractor's work, owner delays resulting in changes in the weather season during which the work is to be performed, changes due to actions or inactions of other trades working on the project for which the owner is responsible, and "constructive changes."

A directed change order or a constructive change typically entitles the contractor to a time extension if the changed work is on the then critical path, and to additional compensation not only for all direct costs, time-related costs, and costs for direct disruption that are caused by the change, but also indirect disruption costs for the unforeseen impact of the change on unchanged work.

The disruptive effect of a change is a function of the size (man-hours and cost) of the change, the nature or scope of the change, the number of changes (although the number of changes may not be a sufficient determining factor in an assessment of cumulative impact), and the impact of the change on the other work. Also critical to the magnitude of the disruptive impact of a change is the time within the engineering and construction cycle when the change is issued. The further into the construction phase of the project, the greater the disruptive impact. If the changes are significant in scope and require significant additional man-hours to perform the changed and/or impacted work, direct and indirect disruption may occur.

The cost of direct disruption that is known and foreseeable should be included in the contractor's change order requests as they are submitted to the owner for approval. The indirect disruption is often unforeseeable and referred to as the cumulative impact of changes. If requests for additional compensation for these indirect disruption costs are not included as part of the change order process because they are not foreseeable, cumulative impact claims may be submitted by the contractor, usually near or shortly after the completion of the project. These cumulative impact claims most often seek recovery of the contractor's additional expenditure of resources, typically labor costs.

By any measure, it is difficult for a contractor to recover claimed costs that allegedly result from the cumulative impact of changes, either during the project, through a request for equitable adjustment and claim negotiations, or through arbitration/litigation. The construction industry, courts, and arbitration panels in the United States generally agree that the theory of cumulative impact is reasonable, and that multiple change orders and other types of delays and disruption can negatively impact the contractor's performance of unchanged work such that a contractor expends additional time, man-hours, and costs in completing its "unchanged" base scope of work. Yet, as will be discussed, the standard of proof set by the courts in proving these cumulative impact claims is burdensome, and their decisions are somewhat subjective. Further, the construction industry has no definitive standard to calculate loss of productivity claims that allegedly result from the cumulative impact of changes. Finally, the concept of cumulative impact claims has not always been accepted in dispute resolution venues outside of the United States.

Cumulative impacts remain largely an ill-defined concept. A more thorough understanding of cumulative impacts as defined by the construction industry and courts and boards will aid the contractor in preparing its damages and proving causation. The information herein provides a blueprint for the contractor seeking to recover costs that result from disruption and the cumulative impact of changes. Conversely, information is also provided that can be used by the owner to identify weaknesses in the contractor's claim submittal, and to better defend against such a claim.

The term "contractor" is used throughout this book to indicate the party claiming damages for disruption and cumulative impact. Subcontractors may also be claimants. The term "owner" is used throughout this paper to indicate the party defending against a disruption and cumulative impact claim. Engineering and construction firms, prime contractors, or construction managers may also be defendants against disruption and cumulative impact claims that are submitted by subcontractors.

Chapter 1 discusses disruption and cumulative impact as defined by the construction industry, as well as by courts and boards. Legal considerations affecting disruption and cumulative impact claims, and the challenges that contractors may encounter to sustain a cumulative impact claim, are discussed in Chapter 2. Methods for estimating loss of productivity man-hours for disruption and cumulative impact claims are presented in Chapter 3, including references to commonly referenced industry studies. Chapter 4 presents examples of productivity loss and cumulative impact calculations using various industry studies and methods. A discussion of quantum/damages quantification methods associated with construction claims is presented in Chapter 5, and these quantum/damages calculation methods are discussed in the context of preparing cumulative impact and other disruption claims. Chapter 6 provides information for preparing a cumulative impact or disruption claim during the project. Finally, a discussion of cause-effect analysis is provided in Chapter 7.

Appendix A includes 18 cases in which a court or board awarded a contractor monies pursuant to its disruption claims. In six of these cases, the theory and legal precedent associated with cumulative impact was specifically discussed. Appendix B includes 31 cases in which a court or board denied a contractor's disruption claims, along with a brief description of the reasons for rejecting the claim. In 24 of these cases, the theory and/or legal precedent associated with cumulative impact was specifically discussed. Appendix C provides a list of documentation that is typically preserved on an engineering and construction project.

This book contains information from numerous published sources, and in many cases, the claims made by the various writers of those publications are restated herein. Therefore, the views and conclusions in this book are not necessarily those of the authors.

Richard J. Long, P.E.

Rod C. Carter, CCP, PSP

Harold E. Buddemeyer

About The Authors

Richard J. Long, P.E., is Founder and CEO of Long International, Inc., one of the world's largest construction claims consulting companies, which also provides project management consulting services. Based in Colorado with offices throughout the U.S. and the Middle East, Long International focuses its practice on owners, engineering and construction firms, and contractors performing oil & gas, petroleum refining, petrochemical, chemical, power, mineral processing, manufacturing, industrial, building, and infrastructure projects worldwide. Mr. Long has over 40 years of U.S. and international engineering, construction, and management consulting experience involving construction contract disputes analysis and resolution, arbitration and litigation support and expert testimony, project management, engineering and construction management, cost and schedule control, and process engineering. As an internationally recognized expert in the analysis and resolution of complex construction disputes for over 30 years, Mr. Long has served as the lead expert on over 300 projects having claims ranging in size from US $100,000 to over US $2 billion. He has presented and published numerous articles and training seminars on the subjects of claims analysis, entitlement issues, CPM schedule and damages analyses, and claims prevention.

Before forming Long International, Mr. Long was Senior Vice President, Contract Administration for a major electrical and mechanical contractor. In this role, he had corporate-wide responsibility for technical management and oversight of the preparation and resolution of construction claims. In addition, he was responsible for the development, training, and implementation of project management policies and procedures to ensure that profit, cost, schedule, scope, quality, and safety objectives were achieved. For 13 years, Mr. Long managed the construction claims practices of two large consulting firms. Prior to his consulting career, Mr. Long gained 13 years of project management and process engineering experience on petroleum refining, oil shale, synfuels, mining, and power generation projects with Tosco, Fluor, and Conoco.

Mr. Long earned a B.S. in Chemical Engineering from the University of Pittsburgh in 1970 and an M.S. in Chemical and Petroleum Refining Engineering from the Colorado School of Mines in 1974. He is a Registered Professional Engineer in the State of Colorado. Mr. Long is based in Littleton, Colorado, and can be contacted at rlong@long-intl.com and (303) 972-2443.

Rod C. Carter, CCP, PSP, is a Principal with Long International and has over 15 years of experience in construction project controls, contract disputes and resolution, mediation/arbitration support, and litigation support for expert testimony. He has experience in entitlement, schedule, and damages analyses on over thirty construction disputes ranging in value from US $100,000 to over US $2 billion. His experience includes heavy civil, nuclear, environmental, chemical, power, industrial, commercial, and residential construction. He is proficient in the use of Primavera Project Planner software, and has extensive experience in assessing the schedule impact of RFIs, change orders, and other events to engineering and construction works.

Mr. Carter specializes in loss of productivity, cumulative impact, and quantum calculations, and has held a lead role in assessing damages on more than a dozen major disputes. In addition, Mr. Carter has developed cost and schedule risk analysis models using Monte Carlo simulations to address the uncertainty of estimates and claims. He has testified as an expert in construction scheduling and damages, and has presented expert findings to an international arbitral tribunal.

Mr. Carter earned a B.S. in Civil Engineering from the University of Colorado at Boulder in 1996, with an emphasis in Structural Engineering and Construction Management. Mr. Carter is based in Littleton, Colorado, and can be contacted at rcarter@long-intl.com and (303) 463-5587.

Harold E. Buddemeyer was a Senior Principal for Long International and passed in January 2022. He had nearly 50 years of experience in all aspects of program and construction project management and construction disputes. His experience included construction and property damage/business interruption claims analysis, preparation, defense, and negotiation of settlements on projects including refineries, offshore oil and gas, petrochemical plants, heavy civil and mining projects, oil sands facilities, nuclear, coal and gas-fired power plants, and building projects. Mr. Buddemeyer's project experience included project cost/schedule control, systems and procedures development and implementation, program planning, as well as capital and operating cost estimating and economic analysis during the design, construction, and start-up phases of a diverse cross section of projects.

Mr. Buddemeyer had over 34 years of construction contract disputes consulting experience. In this regard, he was responsible for entitlement and issue analysis; change order analysis; labor productivity analysis; cost and damages analysis; schedule delay and impact analysis; claim report preparation and rebuttal; negotiation and mediation assistance; the organization, development and maintenance of document databases; assistance to counsel during discovery; and depositions and interrogatory preparation.

Mr. Buddemeyer testified in U.S. and international arbitration. He was enrolled in a professional degree program, majored in applied mathematics and operations research, and minored in Chemical and Petroleum Refining Engineering at the Colorado School of Mines from 1965 to 1970. For further information, please contact Long International's corporate office at (303) 972-2443.

Douglas J. Nutter, Long International's Manager of Graphic Services, prepared the graphical illustrations throughout this text and formatted the text for layout and indexing. He has nearly 40 years of experience in design, illustration, cartography, and graphic production for technical support, arbitration, and litigation. Mr. Nutter is skilled in developing complex technical issues and data into concise presentations for use by counsel, juries, and arbitration panels. Prior to his consulting career, Mr. Nutter gained 12 years of engineering drafting and graphics experience on petroleum refining, oil shale, synfuels, mining, and power generation projects with Tosco Corporation.

Mr. Nutter has provided graphic design, illustration, and technical support for construction claim preparation, analysis, defense, and negotiation of settlements for various parties, including owners, contractors, transit agencies, universities, sureties, financial institutions, law firms, and architectural firms. Mr. Nutter is based in Littleton, Colorado, and can be contacted at dnutter@long-intl.com and (303) 427-4368.

Acknowledgements

The authors make the following acknowledgements regarding the referenced material contained in this publication:

All included and referenced AACE International content is reprinted with the permission of AACE International, 1265 Suncrest Towne Centre Drive, Morgantown, WV 26505 USA. Telephone: 800-858-COST/304-296-8444. Fax: 304-291-5728. www.aacei.org. E-mail: info@aacei.org. Copyright © 2013 by AACE International; all rights reserved.

All included and referenced ABA content is reprinted with the permission of the American Bar Association, 321 N. Clark Street, 20th Floor, Chicago, IL 60654 USA. Telephone: 800-285-2221. www.americanbar.org.

All included and referenced content regarding ASCE is reprinted with the permission of the American Society of Civil Engineers, 1801 Alexander Bell Drive, Reston, VA 20191 USA. Telephone: (800) 548-2723. www.asce.org.

All included and referenced Business Roundtable content is reprinted with the permission of the Business Roundtable, 300 New Jersey Avenue, NW, Suite 800, Washington, D.C. USA. Telephone: 202-872-1260: www.businessroundtable.org.

All included and referenced content regarding the article "Impact of Change Orders on Construction Productivity" is reprinted with the permission of the Canadian Journal of Civil Engineering, Canadian Science Publishing (NRC Research Press), 65 Auriga Drive, Suite 203, Ottawa, ON K2E 7W6 Canada. www.nrcresearchpress.com.

All included and referenced CII content is reprinted with the permission of the Construction Industry Institute, The University of Texas at Austin, 3925 W. Braker Lane. (R4500), Austin, Texas 78759 USA. Telephone: 512-232-3000. www.construction-institute.org.

All included and referenced ELECTRI International content is reprinted with the permission of ELECTRI International, 3 Bethesda Metro Center, Suite 1100, Bethesda, MD 20814 USA. Telephone: 301-215-4539. www.electri.org.

All included and referenced content regarding Charles A. Leonard's M.S. Thesis, "The Effects of Change Orders on Productivity," Concordia University, Montreal, Quebec, 1988, is reprinted with the permission of Charles A. Leonard.

All included and referenced MCAA content is reprinted with the permission of Mechanical Contractors Association of America, 1385 Piccard Drive, Rockville, MD 20850-4340 USA. Telephone: 301-869-5800. www.mcaa.org.

All included and referenced NECA content is reprinted with the permission of the National Electrical Contractors Association, 3 Bethesda Metro Center, Suite 1100, Bethesda, MD 20814 USA. Telephone: (301) 657-3110. www.necanet.org.

All included and referenced content regarding the paper "Loss of Construction Labor Productivity Due to Inefficiencies and Disruptions: the Weight of Expert Opinion" is reprinted with the permission of The Thomas D. Larson Pennsylvania Transportation Institute, The Pennsylvania State University, 201 Transportation Research Building, University Park, PA 16802-4710 USA. Telephone: 814-865-1891. www.larson.psu.edu.

All included and referenced content regarding articles written by Gerald McIniry is reprinted with the permission of Revay and Associates Limited, 4333 St. Catherine Street West, Suite 500, Montréal, Quebec H3Z 1P9 Canada. www.revay.com.

All included and referenced Society of Construction Law material from its Delay and Disruption Protocol is reprinted with permission of the Society of Construction Law, The Cottage, Bullfurlong Lane, Burbage, Hinckley LE10 2HQ. Telephone: 07730 474074. www.scl.org.uk.

Cumulative Impact and Other Disruption Claims in Construction

Richard J. Long, P.E.
Rod C. Carter, CCP, PSP
Harold E. Buddemeyer

LONG INTERNATIONAL

Chapter 1

Disruption and Cumulative Impact Defined

Chapter 1

Disruption and Cumulative Impact Defined

This chapter is intended to provide definitions and background on the concepts of disruption, direct disruption, indirect disruption, and cumulative impact. Section 1.1 considers disruption, the differences between disruption and delay, and the components of disruption. Section 1.2 provides definitions and background on the cumulative impact of multiple changes. Section 1.3 provides further definitions and comparisons of direct disruption and cumulative impact, as well the issues of foreseeable impacts and hidden change order costs.

1.1 Disruption

Several U.S. and international construction industry sources have defined "disruption" as follows:

> *Disruption: (1) An interference (action or event) to the orderly progress of a project or activity(ies). Disruption has been described as the effect of change on unchanged work which manifests itself primarily as adverse labor productivity impacts. (2) Schedule disruption is any unfavorable change to the schedule that may, but does not necessarily, involve delays to the critical path or delayed project completion. Disruption may include, but is not limited to, duration compression, out-of-sequence work, concurrent operations, stacking of trades, and other acceleration measures.*[1]

> *Disruption of Work. Any time the contractor's planned or actual sequence of operations is altered there is a disruption of the work. Typically, disruptions require an extension of the project schedule, idling of equipment, and may require costly changes in material procurement. Even if the change is deductive and fewer construction operations are required, the contractor may need to replan other construction operations, change construction sequencing, or make other changes in planned or current activities at a substantial added cost.*[2]

---

1 "Forensic Schedule Analysis," AACE International Recommended Practice No. 29R-03, 25 April 2011, Section 4.4 A, p. 122.

2 Hester, Weston T., John A. Kuprenas, and T.C. Chang, "Construction Changes and Change Orders: Their Magnitude and Impact," Construction Industry Institute, SD66, p. 6, October 1991.

The degree of disruption caused by a change or impact is dependent upon when the change or impact occurs. Early changes before construction starts may have a disruptive impact to engineering and procurement, but may have little disruptive impact on the construction work. However, a change or impact that affects the critical path or a long lead procurement item can have a significant disruptive impact, if the subsequent installation work associated with that item must be performed out of sequence or in an inefficient manner.

1.1.1 Disruption as Compared to Delay

In its Delay and Disruption Protocol, the Society of Construction Law provides a further definition of disruption, as well as a distinction between delay and disruption, as follows:

> *Disruption (as distinct from delay) is disturbance, hindrance or interruption to a Contractor's normal working methods, resulting in lower efficiency. If caused by the Employer, it may give rise to a right to compensation either under the contract or as a breach of contract.*
>
> *Disruption is often treated by the construction industry as if it were the same thing as delay. It is commonly spoken of together with delay, as in 'delay and disruption'. Delay and disruption are two separate things. They have their normal everyday meanings. Delay is lateness (eg delayed completion equals late completion). Disruption is loss of productivity, disturbance, hindrance or interruption to progress. In the construction context, disrupted work is work that is carried out less efficiently than it would have been, had it not been for the cause of the disruption.*
>
> *Disruption to construction work may lead to late completion of the work, but not necessarily so. It is possible for work to be disrupted and for the contract still to finish by the contract completion date. In this situation, the Contractor will not have a claim for an EOT, but it may have a claim for the cost of the reduced efficiency of its workforce.*[3]

Similarly, the AACE International, The Authority for Total Cost Management (AACE International) notes that disruption does not necessarily involve delay.[4]

Although disruption may cause delay, and delay may cause disruption, they are different issues and, therefore, must be analyzed separately. For example, in *Ascon*, the court declined a claim for lost productivity during a period of delay:

---

3 "The Society of Construction Law Delay and Disruption Protocol," Oxford, October 2002, Section 1.19.1 through 1.19.3, pp. 31–33.

4 "Forensic Schedule Analysis," AACE International Recommended Practice No. 29R-03, 25 April 2011, Section 4.4, *Delay Mitigation and Constructive Acceleration,* p. 122.

> *[The subcontractor's expert witness's] first item is for labor. He calculates it by assessing the full cost of the labour employed during the period of extension. [Counsel for C] submits that that [sic] this is wrong in principle. I agree; the labour content of the contract works is not necessarily increased at all, let alone proportionately, by an extension of time. There may indeed be labour-related losses by reason of delay, for example because of downtime, loss of productivity, repetition of tasks or other uneconomic working, but it is for [the subcontractor] to establish and quantify specific claims for losses of that kind, if they have been suffered.*[5]

The FIDIC form of contracts also discusses that disruption and its associated productivity loss can occur with or without a corresponding delay to a critical activity.[6]

1.1.2 The Components of Disruption

There is general agreement in U.S. construction and legal arenas that the disruption associated with changes to the work has two components: (i) the direct impact of a change; and (ii) the indirect or cumulative impact of that change. For example, the Mechanical Contractors Association of America (MCAA), in "Change Orders, Productivity, Overtime—A Primer for the Construction Industry," states that the contractor must consider direct impacts and cumulative impacts resulting from changes in scope as separate items:

> *It is clear that a contractor must consider both the direct impacts of a loss of labor productivity caused by a change to the contract scope of work, as well as the cumulative impact of changes in scope to the unchanged work.*[7]

In *Centex Bateson*, the board differentiated between direct disruption and cumulative disruption:

> *"Local [or direct] disruption" refers to the direct impact that changed work has on other unchanged work going on around it. Conceptually, for purposes of this appeal "cumulative disruption" is the disruption which occurs between two or more change orders and basic*

5 *Ascon Contracting Ltd. v. Alfred McAlpine Construction Isle of Man Ltd* (1999) 66 Con LR 119 at paragraph 39. See Pickavance, Keith, *Delay and Disruption in Construction Contracts,* 3rd ed., T&F Informa (UK) Ltd, London, 2005, p. 656, paragraph 17.2.

6 See Bunni, Nael G., "The FIDIC Forms of Contract," Blackwell Publishing, Third Edition, 2005, Section 17.6.2.

7 "Change Orders Productivity Overtime – A Primer for the Construction Industry," MCAA, 2012, p. 83.

> *work and is exclusive of that local disruption that can be ascribed to a specific change. It is the synergistic effect...of changes on the unchanged work and on other changes.*[8]

The term "direct impact" (or "direct disruption") refers to costs that are the direct and foreseeable consequences of a change. The MCAA cites judicial authority which defines direct impact as follows:

> *Direct impact is generally characterized as the immediate and direct disruption resulting from a change that lowers productivity in the performance of the changed or unchanged work. Direct impact is considered foreseeable and the disrupting relationship to unchanged work can be related in time and space to a specific change.*[9]

The definition and underlying theories associated with cumulative impact are further addressed below.

1.2 Cumulative Impact Defined

The cumulative impact of multiple changes is a phrase referenced in industry publications and published case law, primarily from the U.S. The terms "cumulative effect" and "cumulative disruption" are also used synonymously with "cumulative impact." Generally, in this text, the term "cumulative impact" has been used.

The Construction Industry Institute (CII) has explained the concept of cumulative impact as follows:

> *When there are multiple changes on a project and they act in sequence or concurrently, there is a compounding effect - this is the most damaging consequence for a project and the most difficult to understand and manage. The net effect of the individual changes is much greater than a sum of the individual parts.*[10]

---

8 *Centex Bateson Constr. Co.*, VABCA Nos. 4613, 5162, 5165, 99-1 BCA ¶ 30,153 (1998), aff'd, *Centex Bateson Constr. Co. v. West*, 250 F.3rd 761 (Fed Cir. 2000) at 149,258 (citing *Triple "A" South*, 94-3 BCA ¶ 27,194, ASBCA No. 46,866 (1994) at 135,523).

9 "Change Orders Productivity Overtime – A Primer for the Construction Industry," MCAA, 2012, p. 82, citing *Centex Bateson Constr. Co.*, VABCA Nos. 4613, 5162, 5165, 99-1 BCA ¶ 30,153.

10 Hester, Weston T., John A. Kuprenas, and T.C. Chang, "Construction Changes and Change Orders: Their Magnitude and Impact," Construction Industry Institute, SD66, p. 35, October 1991.

1.2 – Cumulative Impact Defined

Cumulative impact is typically described as being unforeseeable,[11] as in the following definition from *Centex Bateson*:

> *Cumulative impact is the unforeseeable disruption of productivity resulting from the "synergistic" effect of an undifferentiated group of changes. Cumulative impact is referred to as the "ripple effect" of changes on unchanged work that causes a decrease in productivity and is not analyzed in terms of spatial or temporal relationships. This phenomenon arises at the point the ripples caused by an indivisible body on two or more changes on the pond of a construction project sufficiently overlap and disturb the surface such that entitlement to recover additional costs resulting from the turbulence spontaneously erupts. This overlapping of the ripples is also described as the "synergistic effect" of accumulated changes. This effect is unforeseeable and indirect. Cumulative impact has also been described in terms of the fundamental alteration of the parties' bargain resulting from changes.*[12]

In *McMillan Brothers*, the board provided a similar definition:

> *Because contractors are required to include to include known and generally foreseeable impacts on unchanged work in pricing the cost of a change, the term "cumulative impact" has come to mean, in a generic sense, the impact on unchanged work which is not attributable to any one change but flows from the synergy of the number and scope of changes issue on a project. The underlying theory is that numerous changes cause a cascading ripple-type of impact on performance time and efficiency which is too uncertain or diffuse to be readily discernable [sic] at the time of pricing each individual change.*[13]

Another court defined cumulative impact costs as follows:

> *"Cumulative impact" costs...are costs associated with impact on distant work, and are not readily foreseeable or, if foreseeable, as not [sic] readily computable as direct impact costs. The source of such costs is the sheer number and scope of the changes to the contract. The*

---

11 Although defined as unforeseeable, there may be a time during the project when an experienced contractor will recognize – by i) keeping productivity data on the project; ii) recognizing that its costs are significantly higher than expected; iii) receiving claims from subcontractors for higher man-hour costs; or iv) being cognizant of industry studies on the topic of cumulative impact – that the cumulative impact is no longer unforeseeable and that the project is, in fact, experiencing increased costs as a result of the cumulative impact of change orders and other impacting events.

12 *Centex Bateson Constr. Co.*, VABCA Nos. 4613, 5162, 5165, 99-1 BCA ¶ 30,153 (1998), aff'd, *Centex Bateson Constr. Co. v. West*, 250 F.3rd 761 (Fed Cir. 2000) at 149,258.

13 *McMillin Brothers Constructors, Inc.*, 91-1 BCA ¶ 23,351, EBCA No. 328-10-84, 1990 WL 140900 at p. 12 (citing *Fruehauf Corp.*, PSBCA No. 477, 74-1 BCA ¶ 10,596 (1974) and *Bechtel National*, NASA BCA No. 1186-7, 90-1 BCA ¶ 22,549 (1989).

> *result is an unanticipated loss of efficiency and productivity which increases the contractor's performance costs and usually extends his stay on the job.*[14]

Two important points can be drawn from these definitions. First, cumulative impact is unforeseeable at the time of pricing a given change, and second, cumulative impact cannot necessarily be analyzed by "spatial" (*i.e.*, proximity to changes) or "temporal" (*i.e.*, chronological or sequential) relationships. The later point seems to be echoed by one commentator who stated that cumulative impact occurs where work on one activity is adversely affected by another activity, or by the site environment.[15]

Based on these definitions, the "ripple" or "synergistic effect" from an activity disrupted by several changes does not only impact nearby or follow-on work, but also can potentially impact any and all project work being performed concurrently or subsequent to that activity. It is the effect, the unforeseeable impact, on other base scope work that is at issue with cumulative impact claims.

Loss of productivity can be estimated using industry studies which attempt to correlate the magnitude of project changes with overall project productivity losses. Claims based upon these studies are often referred to as "cumulative impact claims," and the studies themselves refer to cumulative impact.[16] It should be noted that these studies quantify the effect of changes on overall project productivity and, therefore, depending on the circumstances, the lost man-hours estimated using these studies may include both direct disruption and cumulative impact.[17] Therefore, cumulative impact is sometimes used to refer generally to the impact of multiple changes, without

---

14 *Haas & Haynie Corp.*, GSBCA Nos. 5530, 6224, 6638, 6919-20, 84-2 BCA ¶ 17,446 (1984) at 86,897; see also *David H. Tierney Jr.*, GSBCA Nos. 7107, 6198, 5585, 88-2 BCA ¶ 20,806 (1988) at 105,170; *Pittman Construction Co.*, GSBCA Nos. 4,897, 4,923, 81-1 BCA ¶ 14,847, 1980 WL 2643 at 73,297-300.

15 See Robert F. Cushman, Stephen D. Butler, & James F. Nagle, *Construction Change Order Claims,* Wiley Law Publications, § 4.11 (1998 Supplement).

16 By way of example, refer to the titles of these studies: McEniry, Gerald, "The Cumulative Effect of Change Orders on Labour Productivity – The Leonard Study 'Reloaded'," The Revay Report, Volume 26, No. 1, May 2007; Hanna, Awad S., "Quantifying the Cumulative Impact of Change Orders for Electrical and Mechanical Contractors," Construction Industry Institute Research Report 158-11, February 2001; and Ibbs, William and Gerald McEniry, CFCC PSP, "Evaluating the Cumulative Impact of Changes on Labor Productivity – an Evolving Discussion," Cost Engineering, AACE International, Vol. 50/No. 12, December 2008.

17 The overriding issue is that one of the inputs into the studies' formulae is approved change order man-hours, when such man-hours may or may not include all of the direct disruption man-hours. This issue is addressed further in Chapter 3, Section 3.4.1.3.7 herein.

differentiating between direct disruption and cumulative disruption per the definitions provided above.

Certain definitions of cumulative impact refer to the impact of "change orders," as in changes of scope approved by the owner. Other definitions refer to the impact of "changes," a broader and more far-reaching term which does not imply formal owner approval. As added confusion, industry studies which correlate the extent of project change with overall productivity typically use approved change order man-hours to measure loss of productivity.[18] One could ask whether it is only approved change orders that make up a cumulative impact claim.

Arbitration panels, courts, and boards seem to at least consider the effects of requests for information (RFIs) and other impacts.[19] One commentator has written that compensable events such as multiple change orders, RFIs, differing site conditions, suspensions of work, or other work interruptions may be regarded as causing cumulative impact.[20] Therefore, not only can owner-approved changes be used as part of a cumulative impact claim, but also constructive changes[21] and potentially any other disruptive events for which the owner may be responsible. However, the contractor should still recognize any direct disruption that is caused by these events during the project, and submit the required notice and change order documentation as part of a change order request shortly after these events have occurred. Furthermore, because cumulative impact claims can be estimated using the magnitude of the changed man-hours, the contractor should attempt to maintain records demonstrating the direct extra work associated with these events.

18 See Chapter 3, Section 3.4.1.3 herein.

19 See, *e.g., Centex Bateson Constr. Co.*, VABCA Nos. 4613, 5162, 5165, 99-1 BCA ¶ 30,153 (1998), aff'd, *Centex Bateson Constr. Co. v. West*, 250 F.3rd 761 (Fed Cir. 2000) and *Amelco Elec. v. City of Thousand Oaks*, 98 Cal. Rptr. 2d 159 (Cal Ct. App. 2000). See also Appendices A and B herein.

20 See Keating, Geoffrey T. and Burke, Thomas F., "Cumulative Impact Claims: Can They Still Succeed?," p. 4, 20-APR Construction Law, 30, 2000.

21 Constructive changes can include, but are not limited to, defective specifications, over-inspection, changes in the methods of performance, changes in the construction sequence, nondisclosure of information, changes required by regulatory agencies, value engineering, misinterpretation of specifications, incomplete responses to RFIs, and differing site conditions.

1.3 Direct Disruption vs. Cumulative Impact, Foreseeability, and Hidden Costs

One commentator, having evaluated the decisions from U.S. courts and boards, provided the following concise definitions of local disruption and cumulative disruption:

> *Conceived in these terms, local disruption consists of direct, immediate, proximate, and foreseeable consequences based on temporal or spatial relationships flowing out of specific individual change orders.*
>
> *Cumulative disruption, on the other hand, is a combined synergistic effect caused indirectly, unforeseeably, remotely, or distantly by an indivisible group of change orders, is not based on a temporal or spatial relationship and includes all disruption not treated as local disruption.*[22]

Direct or local disruption is understood to include the impact to nearby or follow-on work, and the contractor is expected to recognize and price the resulting costs in the change order.[23] Industry associations have developed methods and studies to provide guidelines for determining the direct disruptive impact of changes.[24] Experienced contractors routinely include foreseeable impacts in their change order pricing, with the understanding that additional compensation for such impacts is foreclosed after the change order has been agreed, as stated below:

> *A contractor is expected to anticipate that it will have to perform all work reasonably foreseeable from the bid documents. If certain impacts to the project could have been foreseen when the contractor and owner entered into the contract, or later when the parties modified the contract, then the contractor is expected to have taken that disruption into account when pricing the contract modifications or change orders. For example, if a solicitation includes contingency options that the owner may choose to award after work on the project has begun, the contractor may not seek to recover for the cumulative impact of the activation of these pre-priced options. In sum, no entitlement to an equitable adjustment exists if the impact to the*

---

22 Finke, Michael R., "Claims for Construction Productivity Losses," 26 Pub. Cont. L.J. 311, p. 319 (Underlining added).

23 See *Haas & Haynie Corp.*, GSBCA Nos. 5530, 6224, 6638, 6919-20, 84-2 BCA ¶ 17,446 (1984) at 86,897.

24 See, for example, the United States Corps of Engineers issued its Modification Impact Evaluation Guide EP 415-1-3, in 1979; Construction Industry Institute issued "Construction Changes and Change Orders: Their Magnitude and Impact," Source Document 66, Oct. 1991; "Estimating Lost Labor Productivity In Construction Claims," AACE International Recommended Practice No. 25R-03, April 13, 2004; and William Schwartzkopf, *Calculating Lost Productivity in Construction Claims,* Chapter 4, 1995. Also, see Chapter 3, Section 3.4.1 herein.

unchanged work was foreseeable at the time the contractor executed the underlying change orders.[25]

Cumulative impact, by contrast, is the unforeseeable "indirect" impact of an undifferentiated group of changes, on other changes and on the base scope work. Contractors often contend that the indirect, disruptive impact of a change cannot be foreseen and measured at the time that the change order is settled. This unforeseeable impact is addressed by many U.S. courts and boards:

> *...costs associated with impact on distant work, and are not as readily foreseeable or, if foreseeable, as readily computable as direct impact costs. The source of such costs is the sheer number and scope of the changes to the contract.*[26]

And:

> *A critical condition precedent to the allowance of cumulative disruption costs ... is showing that they relate to excessive and frequent design or structural changes the impact of which were distant and unforeseeable during the pricing of proposals and negotiations for direct costs.*[27]

Industry sources also recognize the difficulty of properly evaluating the true costs of a change order contemporaneously. In a 2001 research paper, the CII reported on what it identified as the "hidden costs" of changes (which includes lowered productivity) on a project, and the impossibility of estimating such costs contemporaneously:

> *Hidden costs are defined as costs not readily apparent or missed when evaluating project change implementation. A major problem with the execution of project change is failure to consider all the costs associated with implementation. Direct costs such as material, equipment and labor or established indirect costs in the form of overhead are fairly easy to identify and account for in project change estimates. The more difficult task is estimating or predicting the hidden cost associated with change implementation; i.e., delays, lowered productivity, poor communications or rework.*
>
> *The first research objective was to identify and quantify the hidden cost of change. We quickly discovered that it was impossible to accurately estimate all hidden costs associated*

---

25 Jones, Reginald M., "Lost Productivity: Claims for the Cumulative Impact of Multiple Change Orders," Public Contract Law Journal, Volume 31, Number 1, Fall 2001, p. 23, published by the Section of Public Contract Law, The American Bar Association.

26 *Haas & Haynie Corp.*, GSBCA Nos. 5530, 6224, 6638, 6919-20, 84-2 BCA ¶ 17,446 (1984) at 34 (citing *Pittman Construction Co.*, GSBCA Nos. 4,897, 4,923, 81-1 BCA ¶ 14,847, 1980 WL 2643).

27 *Southwest Marine, Inc.*, 94-3 BCA ¶ 27,102, DOTCAB No. 1663, WL 13942 at p. 44.

> *with implementing change prior to change implementation. Even after project change is implemented, it is difficult to capture and account for the "ripple effect"...*[28]

Chapter 6, Sections 6.1 and 6.3, discuss options for estimating cumulative impact during the project.

---

28 Ibbs, C.W. & Walter E. Allen, "Quantitative Impacts of Project Change," CII Source Document 108 at 32, May 1995.

Chapter 2

Legal Considerations

Chapter 2

Legal Considerations

A full discussion of the case law related to cumulative impact claims is not within the scope of this text. Rather, several relevant excerpts from court and board decisions are identified, along with attorney commentary, which should be known to the contractor or expert trying to prepare a supportable cumulative impact claim, or to an owner and its expert defending against such claims.

The Appendices of this text include brief descriptions of 49 cases in which a court or board considered claims or arguments related to the impact of multiple changes. These cases are sourced primarily from U.S. federal courts, the Boards of Contract Appeals from the Veteran Affairs (VABCA), the Armed Services (ASBCA), and the General Services (GSBCA) administrations, as well as certain state courts.

Specifically, Appendix A includes 18 cases in which a court or board awarded a contractor monies pursuant to its disruption claims. In six of these cases, the theory and legal precedent associated with cumulative impact was specifically discussed.

Appendix B includes 31 cases in which a court or board denied a contractor's disruption claims, along with a brief description of the reasons for rejecting the claim. In 24 of these cases, the theory and/or legal precedent associated with cumulative impact was specifically discussed.

In reviewing decisions by U.S. courts and boards, two points become clear: (i) they generally acknowledge the theory of the cumulative impacts of changes; and (ii) they often deny recovery to contractors seeking such claims, for a variety of reasons. These points are further addressed in Sections 2.1 and 2.2 below. Arguments against the theory of cumulative impact are discussed in Section 2.3. In Section 2.4, the *Bell BCI Company v. United States* case is discussed to demonstrate how the contractor prevailed on its cumulative impact claim.

2.1 Acknowledgement of Cumulative Impact

U.S. courts and boards have recognized the theory of the cumulative impacts of changes and have granted recovery to several claimants under this theory.[29] In *Bell BCI Co. v. United States,*[30] the court found that lost man-hours resulted from the cumulative impact of change orders after hearing evidence regarding the contractor's Measured Mile analysis and Earned Value analyses. The court wrote the following:

> *Bilateral modifications agreed to by the parties generally cover the costs and time of performing the changed work. On a project where the Government issues many change orders, bilateral modifications will compensate the contractor for the cost of performing the changed work, but the cumulative effect of the changes may add to the contractor's time and effort in performing the unchanged work as well...*[31]
>
> *...The combination of these conflicting factors created a classic environment for cumulative impact and labor inefficiency. Multiple change orders on a construction project potentially can be accommodated if the owner acknowledges that additional time and money will be required, and if the parties carefully plan the sequencing of the changed work. However, if the owner as here denies the additional time or money to perform changed work, but nevertheless continues the flow of change orders to the contractor, a chaotic project inevitably will result...*[32]

Another board defended the theory of cumulative impact, writing:

> *In effect the...argument is that the effect from all the changes cannot be greater than the sum of all the parts thereof. We do not agree. The entire basis for recovery of an impact or ripple claim is that the effect of changes can and does create costs beyond those attributable to the changes themselves.*[33]

Even when courts or boards deny the contactor's claims, it is typically not due to rejection of the theory of cumulative impact.[34] In *McMillan Brothers Constructors*, for example, the board rejected the contractor's cumulative impact claim for many

---

29 See, *e.g.*, *David H. Tierney Jr.*, GSBCA Nos. 7107, 6198, 5585, 88-2 BCA ¶ 20,806 (1988) and *Bechtel National*, NASA BCA No. 1186-7, 90-1 BCA ¶ 22,549 (1989). See also Appendix A herein.

30 See *Bell BCI Co. v. United States*, No. 03-1613C, 81 Fed. Cl. 617, 2008 U.S. Claims LEXIS 116 (April 2008).

31 *Id.* at p. 29.

32 *Id.* at p. 31.

33 *Ingalls Shipbuilding Division, Litton Systems, Inc.*, ASBCA 17579, 78-1 BCA, ¶ 13,038 (1978) at 63,663.

34 See Appendix B herein.

reasons, but stated that *"we do not question that such impacts from cumulative changes do, in some instances, occur and should be compensated."*[35]

There is general agreement in legal circles in the United States that the cumulative impact of changes is a legitimate cause of action, but the contractor must clear several difficult hurdles to prove and support such a claim.

2.2 Disruption and Cumulative Impact Claim Challenges

To sustain its burden of proof on either a disruption or cumulative impact claim, the contractor must provide compelling evidence. For example, the contractor generally has to present and demonstrate the following to recover costs on a disruption and cumulative impact claim:

- Identify the impacted work and the additional cost of performing that work;
- Identify the events or actions/inactions that caused the impacts, and demonstrate that the increased costs were the result of the events or actions/inactions and why the owner is responsible for the events or actions/inactions;
- Demonstrate that the contractor has reserved its right to recover its increased costs resulting from the events or actions/inactions;
- Properly quantify the disruption and cumulative impact costs. Compare the actual costs and man-hours to the budget baseline costs and man-hours, plus the costs and man-hours that would have been incurred had the actions/inactions not occurred; and
- Demonstrate that the contractor took reasonable steps to mitigate its loss.

If the contractor does not properly address these challenges, either during the project or retrospectively while preparing its claims, then the contractor's disruption and cumulative impact claim may fail. Cumulative impact claims are extremely difficult to prove. Court and board decisions are rife with examples of why these claims are denied. To make matters worse, as one commentator has written, *"courts and boards are*

35 *McMillin Brothers Constructors, Inc.,* 91-1 BCA ¶ 23,351, EBCA No. 328-10-84, 1990 WL 140900 at p. 12.

not always consistent in their treatment of cumulative impact claims."[36] Still, an understanding of why claims are rejected should help the contractor to avoid similar mistakes.

Challenges that a contractor may need to overcome to prevail on a disruption or cumulative impact claim include the following items:

- Demonstration of causation;
- Reservation of the right to make a claim;
- Use of proper damages methods;
- Preservation of good records; and/or
- The claim premise being barred by law or contract.

The discussions that follow refer to United States and other case law to illustrate the propositions. However, these issues are likely to be considered by any dispute resolution forum or venue irrespective of governing law. For example, causation is a requirement of most legal systems.

2.2.1 Demonstration of Causation

Causation is an extremely important element that must be established in any construction-related claim, and courts and boards often deny cumulative impact claims for lack of causation. As one board wrote, cumulative impact claims "*...are routinely denied because there were an insufficient number of changes, contractor-caused concurrent delays, disruptions and inefficiencies and/or a general absence of evidence of causation and impact.*"[37]

The contractor must first establish that the changes were excessive. Analysts agree that the quantity of changes alone does not establish a claim's validity, but that a combination of the quantity and the total cost (man-hours) of the change orders compared to the base contract may be more indicative that a cumulative impact has occurred.[38]

---

36 Jones, Reginald M., "Lost Productivity: Claims for the Cumulative Impact of Multiple Change Orders," Public Contract Law Journal, Volume 31, Number 1, Fall 2001, p. 45, published by the Section of Public Contract Law, The American Bar Association.

37 *J. A. Jones Construction Co.,* 00-2 BCA ¶ 31,000, ENGBCA Nos. 6390-1, 6386-7, 6348, 6388-9, 2000 WL 1014011 at p. 47.

38 See Jones, Reginald M., "Lost Productivity: Claims for the Cumulative Impact of Multiple Change Orders," Public Contract Law Journal, Volume 31, Number 1, Fall 2001, pp. 23–24, published by the Section of Public Contract Law, The American Bar Association; Backus, Leland E., "The Cumulative Impact Claim: What Does it Take to Prove?," ABA Forum on the Construction Industry, 2001; and Sgarlata, Barnes & Brasco, "Successful Project Management through an Understanding of the Risks and Benefits of Early Claims Resolution," (CMAA 2004 National Conference).

However, it is clear that causation must first be established.

In *Freeman-Darling, Inc.*,[39] the occurrence of numerous change orders amounting to 10 percent of the original contract price did not persuade the court that a substantial cumulative effect on the contractor's work had occurred.

In *Pittman*, the United States General Services Board of Contract Appeals (GSBCA) denied the contractor's cumulative impact claim where 206 changes were issued, amounting to approximately $3 million on a base contract of $25 million. After a review of the change orders in question, the board found that the scope of the contract was essentially unchanged, and that the contractor had in practice been required to do no more than build the facility required under the contract as originally awarded.[40] The *Pittman* board also concluded that the contractor had itself deviated from the planned construction sequence. Consequently, where both parties contributed to the delay, neither party could recover unless there was clear proof of apportionment.

Although multiple change orders may have been issued on a project, courts may not accept a cumulative impact claim on the mere proof that there were numerous change orders.[41] In *Wunderlich Contracting Co. v. United States*,[42] the court held that the 35 change orders amounting to 6 percent of the contract value did not prove that the government was responsible for all the losses on the contract.

The *Wunderlich* court stated:

Construction industry studies estimate the effect of cumulative impact based upon the amount of change order man-hours, not change order costs. A change order revising the type or size of a compressor to be installed may have a substantial cost, but no impact on labor productivity, if the man-hours for installing the revised type of compressor are not much different than the man-hours for the original compressor. Also see *Hensel Phelps*, ASBCA No. 49270, 99-2 BCA ¶ 30,531 at 150,795 [citing *Santa Fe Eng'g, Inc.*, ASBCA Nos. 24578, 25838, 2838, 28687, 94-2 BCA ¶ 26,872, where the court found that "*the number of RFIs and changes alone is insufficient to establish the Government's liability for a contractor's inefficiency*."

39 See *Freeman-Darling, Inc.*, GSBCA No. 7112, 89-2 BCA ¶ 21,882 (1989).

40 See *Pittman Construction Co.*, GSBCA Nos. 4,897, 4,923, 81-1 BCA ¶ 14,847, 1980 WL 2643.

41 See Jones, Reginald M., "Lost Productivity: Claims for the Cumulative Impact of Multiple Change Orders," Public Contract Law Journal, Volume 31, Number 1, Fall 2001, pp. 21–23, published by the Section of Public Contract Law, The American Bar Association.

42 See *Wunderlich Contracting Co. v. U.S.*, 173 Ct.Cl. 180, 351 F.2d 956 (1965).

> *There is no exact formula for determining the point at which a single change or a series of changes must be considered to be beyond the scope of the contract and necessarily in breach of it. Each case must be analyzed on its own facts and in light of its own circumstances, giving just consideration to the magnitude and quality of the changes ordered and their cumulative effect on the project as a whole.*[43]

The court reasoned that even though the contractor's performance had been lengthier and costlier than anticipated at the time that its bid was submitted, the contractor ultimately constructed essentially the same project that was agreed upon in the contract. Therefore, the changes neither materially altered the nature of the bargain, nor were they so beyond the original scope as to constitute a breach of contract; that is, "Cardinal Change."[44] The court also held that there was a critical lack of proof of the cause-effect relationship between the events and the claimed impact to labor productivity and other costs.[45]

In *Southwest Marine, Inc.*,[46] the United States Department of Transportation Board of Contract Appeals (DOTBCA) denied the contractor's cumulative impact claim because the contractor failed to (i) show that the work differed significantly from the work anticipated under the original contract, even though 202 change orders had been issued, and (ii) present evidence sufficient to demonstrate that the change orders had actually caused Southwest Marine to experience labor inefficiency and other disruptive effects.[47]

Courts and boards have not set out threshold values with respect to the quantity and relative value of change orders below which claims are denied, and above which they are successful. This type of analysis is too simplistic. Thus, in *Pittman*, the contractor's claim was not denied because the change order value was only 12 percent of the base contract, and in *Southwest Marine, Inc.*, the contractor's claim was not denied despite it having experienced 202 change orders. In both cases, the claims were denied because the contractors had failed to prove that the changes had severely impacted their productivity.[48] This is the fundamental issue.

---

43 *Id.* at p. 13 (Westlaw print).

44 See Bramble, Barry B. and Michael T. Callahan, *Construction Delay Claims*, Third Edition, 2000, Section 5.05 [H].

45 *Id.* at Section 5.06 [B].

46 See *Southwest Marine, Inc.*, DOT BCA No. 1663, 94-3 B.C.A. (CCH) ¶ 27,102 (1994).

47 See Bramble, Barry B. and Michael T. Callahan, *Construction Delay Claims,* Third Edition, 2000, Section 5.05 [H].

48 Accordingly, an evaluation of the existing case law by reference to the value of change is unlikely to be meaningful, and the more important aspect is to determine the quality of proof.

Courts and boards acknowledge the difficulty of demonstrating the link between changes and loss of productivity. As one board stated, *"the conceptual problem is that there is no easy way of drawing a line between those costs that directly result from changes and those that are more remote."*[49] In *Centex Bateson*, the VABCA, in rejecting a contractor's cumulative impact claim, was more specific:

> *To find entitlement to cumulative impact damages here, we would have to find that an event involving changing the location of a light switch in a room in the east basement of VAMC Houston impacted the efficiency of all its work on the west side of the sixth floor simply because Dynalectric overran its labor projections for certain work activities for that month. The record simply does not support such a finding.*[50]

This excerpt highlights the contractor's burden in proving causation. Showing a large number of changes and a corresponding cost overrun is typically not enough. Some meaningful connection or correlation between changes and cost growth must be drawn. Explaining how changing the light switch's location in the east basement can impact work in other parts of the building is not an easy task. Likewise, explaining how adding additional pipe in one process unit can impact the electrical work in another process unit is not an easy task, and it must be remembered that the contractor bears the burden of proof.

In the *Centex* case, the VABCA suggested that proof *"may take the form of demonstrating that there are no other reasons for a loss of productivity for which the Government is not responsible."*[51] This suggestion may not be applicable in every circumstance. The contractor in this case submitted a Total Cost claim, and under this method had to prove that its actions did not negatively impact the work in any way.

When the contractor is itself responsible for some of the disruptions to the work conditions, the owner should not be accountable for such working condition disruptions, particularly if the contractor cannot distinguish and apportion between its own disruptions and disruptions caused by the owner.[52] At the least, the existence of contractor-caused impacts adds an additional element of uncertainty regarding disruption, and increases the margin of error in a cumulative impact estimate.

Likewise, a research team also concludes that causal linkages are necessary because:

---

49 *Pittman Construction Co.*, GSBCA Nos. 4897, 4923, 81-1 BCA ¶ 14,847, 1980 WL 2643 at p. 18.

50 *Centex Bateson Constr. Co.*, VABCA Nos. 4613, 5162, 5165, 99-1 BCA ¶ 30,153 (1998), aff'd, *Centex Bateson Constr. Co. v. West*, 250 F.3rd 761 (Fed Cir. 2000) at p. 85 (Westlaw print).

51 *Id.* at p. 82.

52 *Id.*

> *(1) even when disruptions are initially caused by one party (e.g., the owner), the other party (e.g., the contractor) may be able to reduce or escalate the disruptions and inefficiency throughout the course of work; and (2) productivity losses rarely result from a single causing factor but multiple and concurrent ones for which both parties can be responsible.*[53]

Similarly, the VABCA in *Centex Bateson* found a lack of causation linking events to the contractor's cumulative impact losses, stating as follows:

> *The current case law recognizes that using the "synergy" analysis of cumulative impact necessarily leads to a total cost damage analysis. However, even if we were willing to make the leap of faith and find that the 1,561 events identified by Dynalectric somehow combined into a synergistic, cumulative impact on unchanged work, the claim fails for lack of proof of damages.*[54]

In *Clark Constr. Group, Inc.*, the VABCA also found that the contractor did not sufficiently demonstrate that the problems which it encountered cumulatively impacted the work and reduced labor efficiency:

> *[T]o the extent that [the claim] includes loss of labor productivity caused by the combined effect of the change of sequence, wet conditions and late RFI responses in addition to the alleged direct efficiency losses, [the subcontractor] has not met the test we established in Centex Bateson to show that the combination of the alleged conditions cumulatively impacted the work and reduced labor efficiency.*[55]

In *David H. Tierney*, the GSBCA granted recovery on the contractor's cumulative impact claims, stating:

> *Although we are not able to pinpoint, day by day, the effect of each change on each item of work, we do find that some of those changes had a cumulative impact on job progress as a whole, for which appellant is entitled to compensation.*[56]

Yet the GSBCA in *Centex Bateson*, having reviewed the above decision and other decisions granting recovery of cumulative impact damages, opined that the recovery

---

53 Ibbs, William, Long D. Nguyen, and Seulkee Lee, "Quantified Impacts of Project Change," Journal of Professional Issues in Engineering Education and Practice, ASCE, January 2007, p. 45.

54 *Centex Bateson Constr. Co.*, VABCA Nos. 4613, 5162, 5165, 99-1 BCA ¶ 30,153 (1998), aff'd, *Centex Bateson Constr. Co. v. West*, 250 F.3rd 761 (Fed Cir. 2000) at p. 84 (Westlaw print).

55 *Clark Construction Group, Inc.*, 00-1 BCA ¶ 30,870, VABCA No. 5674 (April 2000) at 152,413.

56 *David H. Tierney Jr.*, GSBCA Nos. 7107, 6198, 5585, 88-2 BCA ¶ 20,806 (1988) at 105,121.

was more the result of *"subjective conclusions rather than hard, objective evidence."*[57] The board further explained:

> *We find little difference between the proof adduced in the cases where recovery was allowed and those where recovery was not allowed. That is, in some cases, the panel concluded that there was sufficient proof of causation due to the number of changes and labor cost overruns and other panels, on the same subjective basis, came to the opposite conclusion.*[58]

The board went on to compare the legal analysis of impact costs with one judge's comments on obscenity, when he stated that he could *"never succeed intelligently"* in defining obscenity but *"I know it when I see it."*[59]

Hence, there is no definitive formula for determining the quantity and magnitude of changes that constitute a basis for recovery on cumulative impact claims, nor is there any established standard for proving how two or more change orders have synergistically impacted other basic work. Courts and boards seem to look at the facts and circumstances of each case independently and, by their own admission, make a subjective decision as to the merit of the claim.

Chapter 7 provides recommendations for proving causation in cumulative impact claims.

2.2.2 Reservation of Rights to Make a Claim

Imprecise wording on change order forms often causes problems relating to what costs are included in the agreed price for the change. Contractors may argue that the change order language, such as compensation for "indirect costs, or time-related costs," does not include loss of productivity or cumulative impact costs. The contractor argues, in defense of its position, that there is no way that it could have known the actual impact of later owner-caused changes at the time that the earlier change order was priced. Therefore, the contractor waits until the end of the job to quantify and request its impact or loss of productivity claim, and calls it a "cumulative impact claim." The owner, however, often refuses to pay any more than the price earlier agreed.[60]

Thus, an owner's primary defense to a disruption and/or cumulative impact claim is often to argue that, in executing the various change orders, the contractor agreed to

---

57 *Centex Bateson Constr. Co.*, VABCA Nos. 4613, 5162, 5165, 99-1 BCA ¶ 30,153 (1998), aff'd, *Centex Bateson Constr. Co. v. West*, 250 F.3rd 761 (Fed Cir. 2000) at p. 82 (Westlaw print).

58 *Id.* at p. 83.

59 *Id.* citing *Jacobellis v. Ohio*, 378 U.S. 184, 197 (1964).

60 Also see Chapter 2, Section 2.2.2 and Chapter 6, Section 6.1 herein.

the price of the changes and thereby waived its rights to seek any further compensation. The argument advanced is that the contractor, by signing the change order, included any possible costs that would be associated with the cumulative impact claim and accepted any risk that the costs would be more than anticipated.

For example, the change order form may contain the following commonly used language:

> *This change includes all costs associated with the scope of work associated with this change, including all direct, indirect, and impact costs on the unchanged work such as loss of productivity, ripple effect, cumulative impact, and acceleration.*

The owner's argument is simply that the contractor, by signing the change order, has accepted the risk associated with the changed work and has included in the change order price the value of any possible costs that could be associated with disruption and/or cumulative impact.[61]

If an impact is foreseeable, it may be reasonable to consider it a direct impact, and any right to claim the costs associated with the disruption is either priced, settled, or waived with the execution of a change order. If an impact is unforeseeable, it may be reasonable to consider it an indirect impact, and the right to recover for an indirect impact should be reserved, or at a minimum, the contractor must not waive the right, so that the possibility of recovery exists.[62] The handling of these issues is highly dependent on the contract and change order form language.

The MCAA acknowledges the difficulty that a contractor will face if it signs a change order with what is termed "full accord and satisfaction"[63] language on the change order form, as follows:

> *Often, contract language known as "full accord and satisfaction" language, contained in some change order forms, may require the contractor to attempt to price all categories of productivity loss within the change itself, as estimated values. This is called a forward priced productivity loss and the cost of this estimated loss can be included as a line item in the change order proposal. While it can be highly beneficial to include all estimated impacts*

---

61 The legal defense is essentially "accord and satisfaction."

62 See Jones, Reginald M., "Lost Productivity: Claims for the Cumulative Impact of Multiple Change Orders," Public Contract Law Journal, Volume 31, Number 1, Fall 2001, pp. 23–24, published by the Section of Public Contract Law, The American Bar Association.

63 Accord and satisfaction is a concept attributed to a bilateral contract, and requires (1) proper subject matter, (2) competent parties, (3) a meeting of the minds, and (4) consideration. See *Jackson Construction Co. , Inc. v. U.S.* (Fed Cl. 2004), Lois Law Federal District Court Opinions, No. 97-31C, at p. 16.

> *within a change order, thus "closing out" the change, many owners refuse to recognize labor productivity impacts caused by scope changes or other factors beyond the control of the contractor. This leaves the contractor in the unwanted position of either not executing change orders due to the risk of waiving its rights or placing a "reservations of rights" statement on each change, which can have the effect of holding open the option of making further requests for equitable adjustment should the contractor suffer productivity losses due to the cumulative impacts of changes in scope on the project.*[64]

The MCAA also cautions the contractor about accord and satisfaction, as follows:

> *While the owner-directed change is easiest to identify, the contractor must consider schedule and productivity impacts in the analysis and pricing of the change. Time and productivity impacts often are not the subject of initial negotiations regarding the scope, but these topics should be incorporated into the negotiation process. Failure to consider these impacts at the time of the change can result in a waiver of a contractor's ability to recover additional time and money. Contractors should consider their need to reserve rights associated with impacts if they are not quantified and included in the change order price at the time of each change order negotiation.*[65]
>
> *Courts and boards have found that such language [accord and satisfaction] may bar the contractor from additional recovery. If there is concern that productivity impacts, cumulative impacts of multiple changes, and delay-related costs cannot be quantified for individual changes, the contractor should consider reserving its right to make a claim for such impacts separate from individual change orders.*[66]
>
> *In some cases, the interpretation of the "full accord and satisfaction" language is so broad that the contractor's cumulative impact claim is barred in total or in part because such impacts are claimed to arise from the change orders containing such exculpatory language.*[67]

While this argument is dependent upon the change order language, and the intent and knowledge of the parties at the time of change order signing, courts and boards have denied many claims on this basis.

For example, in the appeal of *Dyson & Co.*, the board acknowledged that the contractor likely suffered cumulative impacts from change orders, but denied recovery, ruling that the modifications *"represent full settlements for the cost consequences flowing from each of the directed changes, and there is no evidence that appellant, which agreed to these*

---

64 "Change Orders Productivity Overtime – A Primer for the Construction Industry," MCAA, 2012, p. 84.

65 *Id.* at p. 5.

66 *Id.* at p. 13.

67 *Id.* at p. 98.

modifications without manifesting any reservations, considered them otherwise."[68] The board further relied on the fact that the modifications were executed after the changed work was almost completed and, therefore, the contractor had knowledge of the impacts.[69]

In *Pittman,* the board did not definitively state that the contractor's cumulative impact claim was barred by executed change orders, but that the board was *"relatively strict in holding that priced out change orders bar recovery of further costs associated with those changes."*[70] The board also examined project correspondence to determine if Pittman had reserved its rights to assert the cumulative impact claim (prior to agreeing to the change order), and commented:

> *Now Pittman comes back, after the entire matter has been concluded, and says it does not like the result. But it managed to go through 206 change orders and thirteen extensions of time without raising a single objection that sufficed to reserve a claim for further equitable adjustment.*[71]

The board denied Pittman's appeal due to lack of causation.

In *Vanlar Construction, Inc. v. County of Los Angeles,*[72] the owner issued seven supplemental agreements and 81 change orders. The contractor signed each document with general waiver and release language. The court acknowledged that the large number of changes may have had a cumulative impact on the contractor's performance costs, but concluded the right to such a claim had been waived. The court determined that if the contractor contemplated a future claim for impact costs, it should have requested that the change orders contained a reservation clause before it signed the documents.[73] After the change order is signed, the court ruled that it was too late to make a claim for such costs.

In *Central Mechanical Construction,*[74] the contractor attempted to recover additional costs by way of a cumulative impact claim after agreeing to settle multiple changes. The

---

68 *Dyson & Co.,* ASBCA No. 21,673, 78-2 BCA ¶ 13,482, 1978 WL 2368 at pp. 25–26 (Westlaw print).

69 See *Dyson & Co.,* ASBCA No. 21,673, 78-2 BCA ¶ 13,482, 1978 WL 2368.

70 *Pittman Construction Co.,* GSBCA Nos. 4,897, 4,923, 81-1 BCA ¶ 14,847, 1980 WL 2643 at pp. 20–21.

71 *Id.* at p. 45.

72 See Bramble, Barry B. and Michael T. Callahan, *Construction Delay Claims,* Third Edition, 2000, p. 5-41, citing *Vanlar Construction, Inc. v. County of Los Angeles,* 217 Cal. Rptr. 53 (Cal. App. 1985).

73 See Bramble, Barry B. and Michael T. Callahan, *Construction Delay Claims,* Third Edition, 2000, Section 5.06 [C].

74 See *Central Mechanical Construction,* ASBCA No. 29434, 86-3 BCA ¶ 19,240 (1986).

board determined that the waiver and release language in the settled change orders barred further recovery.

In *Atlantic Dry Dock Corp. v. United States*,[75] the court held that the release provision in each modification clearly addressed delay and disruption, stating that the modifications incorporated a final settlement of all claims arising out of each modification *"including all claims for delays and disruptions resulting from, caused by, or incident to such modifications or change orders."* The court found that this language was unambiguous and *"reasonably susceptible to only one interpretation,"* which was that the parties intended to resolve all claims for damages caused by delay and disruption, whether cumulative or otherwise. Therefore, the provision was unambiguous and the contractor's claim was barred for cumulative impact.

More recently, in *Jackson Construction Co., Inc. v. United States*,[76] the contractor executed 24 modifications, each of which contained a release provision stating: *"It is further understood and agreed that this adjustment constitutes compensation in full on behalf of the contractor and his subcontractors and suppliers for all costs and markup directly or indirectly, including extended overhead, attributable to the change order, for all delays related thereto, and for performance of the change within the time frame stated."*[77] The court held that this language was sufficient to constitute an accord and satisfaction, which barred the contractor's cumulative impact claim.

States have also upheld the provisions of change orders which include language constituting final settlement of all claims arising out of the changed work. In *Uhle v. Tarlton Corp.*,[78] the court ruled that because the subcontractor had not added any language to the change orders that specifically preserved its right to cumulative impact costs, the subcontractor had waived them.

Notwithstanding these legal precedents, certain courts have held that unless cumulative impact claims had been specifically discussed or considered as part of the change order negotiations, a release may not apply to a cumulative impact claim.[79] Also, if discussions have occurred during change order negotiations regarding cumulative impact claims, but no settlement was reached, the general waiver and release language may not apply to a subsequently submitted cumulative impact claim.[80] If no release language is included in executed change order documentation, a contractor may be entitled to subsequently claim cumulative impact costs, even if it

75 See *Atlantic Dry Dock Corp. v. United States,* 773 F. Supp. 335, 338-39 (M.D. Fla. 1991).

76 See *Jackson Construction Co. v. United States,* 62 Fed.Cl. 84 (Fed. Cl. 2004).

77 *Id.* at p. 9 (Westlaw print).

78 See *Uhle v. Tarlton Corp.*, 938 S.W.2d 594 (Mo. Ct. App. 1997) .

79 See *David H. Tierney Jr.*, GSBCA Nos. 7107, 6198, 5585, 88-2 BCA ¶ 20,806 (1988).

80 See *Saudi Tarmac Co., Ltd and Tarmac Overseas, Ltd.*, ENGBCA No. 4841, 89-3 BCA ¶ 22,036 (1989).

did not expressly reserve its rights to those costs.[81] If the owner will not execute change orders without a release, the contractor should notify the owner in writing that it may pursue cumulative impact claims in the future.

Also, in *Bell BCI Company v. United States*,[82] the U.S. Federal Court of Claims awarded the contractor a $2 million claim for cumulative impact on behalf of its subcontractors despite release language it negotiated and agreed upon contract modifications that were intended to preclude subsequent contract cost adjustments for work associated by the prior modifications, which were in excess of $21 million. The government issued more than 200 modifications to the contract, which delayed the contract by more than 19 months and increased the contract price by 34 percent.

The court confirmed that:

> *Until 1967, the Rice doctrine Rice v. United States 1942 precluded courts from considering the effect of change orders on portions of the work not directly covered by the change ... To avoid inequitable results, the Government changed the standard Changes clause in late 1967 to add language that covers the effect of changes on unchanged work.*[83]

The court found that the government had breached its implied duty of good faith and fair dealing in the administration of the project:

> *Multiple change orders on a construction project potentially can be accommodated if the owner acknowledges that additional time and money will be required, and if the parties carefully plan the sequencing of the changed work. However, if the owner as here denies the additional time or money to perform changed work, but nevertheless continues the flow of change orders to the contractor, a chaotic project inevitably will result. In this case, there were 279 EWOs and 113 contract modifications issued ... while NIH project personnel were maintaining that no further changes would be issued. The project environment was contentious, as NIH representatives bordered on bad faith in denying payment to Bell for extra work performed.*[84]

The court held that the broad release language agreed in the contract modifications did not bar a cumulative impact claim, because the release language in the contract modifications did not contain any language expressly releasing claims for the cumulative impact of the changes. The court also found that the government failed to present any evidence that the adjustments to the contract sum included any payment specifically attributed to the cumulative impact of the various changes.

---

81 See *Beaty Electric*, EBCA No. 408-3-88, 90-2 BCA (CCH) ¶22,829 (1990).

82 See *Bell BCI Co. v. United States*, No. 03-1613C, 81 Fed. Cl. 617, 2008 U.S. Claims LEXIS 116 (April 2008).

83 *Id.* at p. 29.

84 *Id.* at p. 31.

To avoid a legal challenge of accord and satisfaction, the prudent contractor should reserve its rights to claim for cumulative impacts as soon as the problem becomes apparent. If the owner will not accept such reservations in the change orders, and the contractor believes it must sign the change orders to receive compensation for the work it must perform, the contractor should either attempt to include additional costs in its change order pricing, or send the owner a separate letter reserving its right to claim for cumulative impact costs when the full extent of such impacts are known.

Also, to protect against such claims, the owner should include language in a change order or contract modification explicitly releasing any cumulative impact claim. In addition, even where the contract modification or change order contains a clear and unambiguous release of a cumulative impact claim, the owner should expressly allocate a portion of the contract adjustment to compensation for any cumulative impact of each agreed change.

2.2.3 Use of Proper Damages Methods

Owners often argue that the contractor should have been able to track the changes and related impacts in its cost and man-hour accounting system, and to demonstrate the "actual cost" impact of each change. However, although they should, some contractors do not maintain adequate records to enable such actual cost accounting of change orders and their impacts, and some contractors do not separately track the actual costs of changes to the work. They estimate the cost of the changes and obtain contract cost and time adjustments, but continue to record the actual man-hours and cost of the base scope of work, rather than providing more detailed cost controls to track multiple change orders separately from the original contract scope of work.

In addition, some construction cost accounting and control systems do not provide for periodic productivity reports and, therefore, actual jobsite productivity data must be compiled as a separate exercise. Periodic productivity data includes actual craft labor man-hours and installed quantities measured and recorded by period for various areas and disciplines of the project, which can be difficult to maintain, but is often fundamental to the proof of damages and causation. Recognition that indirect cumulative impact costs have been incurred often only occurs at the end of the project, when it is too late to capture such impact costs separately from the base scope of work costs. In these circumstances, other methods must be utilized to estimate the contractor's cumulative impact costs.

The contractor's expert in *J. A. Jones* used the Measured Mile method to calculate the contractor's compensable productivity loss damages. The board rejected this analysis because it found his method to be flawed. The government's expert criticized the analysis as *"one of a kind,"* and demonstrated that the criteria for selecting un-impacted and impacted work was illogical.[85] While the Measured Mile method is generally considered appropriate for calculating loss of productivity, in this case it was implemented improperly.

Another problem is that contractors often calculate loss of productivity claims using the Measured Mile method for one craft or work activity, such as concrete, and then apply the calculated loss of productivity percentage to all other crafts or work activities. Often these other crafts or work activities have no relationship to the craft or work activity actually analyzed; are not on the same schedule activity path; and are months or even years downstream of the originally impacted work. Also, some of these other crafts are found to be impacted by the contractor's own problems, rather than owner-caused impacts. Therefore, such extrapolation may be entirely inappropriate.

Another often cited but disfavored method is a "total loss," which is sometimes called a "Total Cost" or "global" claim. One author has criticized such claims because they do not establish that the claimed damages flow from the events for which the owner's liability is established and other problems for which the contractor is responsible, such as "bid error."[86] It is an un-particularized claim for reimbursement.

A total loss claim contends that the claimed costs are not caused by the actions or inactions of the contractor and are caused only by the factors for which the owner is responsible, and does not detail separate damages as being the result of each specific event. A global claim, on the other hand, does not separate the liabilities between the owner and contractor, and instead claims that the owner was responsible for all losses. Thus a global claim comprises a total loss claim, but not all total loss claims are global claims.

To succeed in a Total Cost claim, the contractor must establish that:

- *The nature of the [claimant's] particular losses make it impossible or highly impracticable to determine them with a reasonable degree of accuracy;*
- *The claimant's bid or estimate was realistic;*
- *Its [the claimant's] actual costs were reasonable; and*
- *It [the claimant] was not responsible for the added expenses.*[87]

---

85 See *J. A. Jones Construction Co.*, 00-2 BCA ¶ 31,000, ENGBCA Nos. 6390-1, 6386-7, 6348, 6388-9, 2000 WL 1014011.

86 See Pickavance, Keith, *Delay and Disruption in Construction Contracts,* 3rd ed., T&F Informa (UK) Ltd, London, 2005, p. 766, paragraph 19.11.

87 See *WRB Corp. v. United States*, 183 Ct.Cl. 409, 426-427 (1968) at paragraph 8.

In a Scottish case, the court set forth the hurdle that the contractor must clear to prevail on a global claim:

> *The logic of a global claim demands, however, that all the events which contribute to causing the global loss be events for which the defender is liable. If the causal events include events for which the defender bears no liability, the effect of upholding the global claim is to impose on the defender a liability which, in part, is not legally his.*
>
> *That is unjustified. A global claim, as such, must therefore fail if any material contribution to the causation of the global loss is made by a factor or factors for which the defender bears no legal liability. ... The point has on occasion been expressed in terms of a requirement that the pursuer should not himself have been responsible for any factor contributing materially to the global loss, but it is in my view clearly more accurate to say that there must be no material causative factor for which the defender is not liable.*[88]

The Total Cost method for calculating cumulative impact damages is notable in U.S. case law both for its disrepute and its status as a method of last resort. Most cases surveyed decline to accord the Total Cost method much respect. In the words of the *WRB* court: *"This theory has never been favored by the court and has been tolerated only when no other mode was available and when the reliability of the supporting evidence was fully substantiated."*[89] Much of the unreliability of the Total Cost method arises because *"it is extremely difficult to assure that the contract is not transformed into a de facto cost reimbursement contract and that costs which should be borne by [the contractor] are excluded."*[90]

In the *Bechtel National*, the board found that, despite the moniker of *"productivity analysis,"* Bechtel had engaged in Total Cost analysis, and that as a result Bechtel had the task of proving causality:

> *Respondent contends that Appellant has presented a total cost claim. Appellant did not establish the specific effects caused by particular RFIs, but determined the amount of disruption by comparing total manhours expended on the project with the total estimated at the time of Forecast 3, as later adjusted at the time Appellant prepared the claim. Appellant maintains that its claim methodology was the best available method for computing its cumulative impact claim because the use of a factor (e.g. 21.7% loss of efficiency) would be less reliable and the total cost method could not account for other causes of increased costs. We find that Appellant has used a total cost approach. A claim of extra hours worked based on virtually the entire difference between an estimate of hours and the actual hours worked is a typical total cost approach. Appellant must, therefore,*

---

88 *John Doyle Construction Limited v. Laing Management (Scotland) Ltd* [2002] BLR 393 (Outer House) and [2004] BLR 295 (Inner House), paragraph 36.

89 *WRB Corp. v. United States*, 183 Ct.Cl. 409, 426-427 (1968) at paragraph 8.

90 *Id.*

> *establish that the Government caused its increased costs with proof that satisfactorily eliminates other causes of inefficiency or delay.*[91]

Industry commentators have cited a limited number of cases where total loss claims have been accepted.[92]

However, total loss claims have been rejected even when the contractor attempts to reduce the calculated amount of its claim by assigning itself a portion of the damages liability. For example, in *Pittman*, the board rejected the contractor's use of the Total Cost method to calculate its damages. Pittman made a claim on behalf of two subcontractors, F&M and Babst. Both subcontractors calculated its total losses, and applied a reduction factor against these losses in its claim. F&M claimed 75 percent of its total losses; Babst claimed 70 percent of its losses. The board still considered the calculation a Total Cost claim, and dismissed the adjustments as a "discount" to the government and the calculations as "flimsy." It further explained how the Total Cost method was held with *"great disfavor"* by courts and boards.[93]

Regarding a cumulative impact claim that is calculated as a Total Cost claim, the contractor must demonstrate that it made no material oversight in preparing its bid productivity estimate, is blameless for the loss of productivity, and could not have foreseen and addressed cumulative impact in the agreed change orders. This is a significant burden of proof given the degree of the contractor's control of its productivity, the sophistication of today's construction accounting and scheduling systems, and the likelihood that a reasonable number of changes will occur on every construction project.

These problems can seemingly be rectified by using a qualified expert who employs the correct method(s) to calculate the contractor's damages. Chapter 3 discusses various productivity loss calculation methods, and Chapter 4 presents sample productivity loss calculations in detail.

---

91 *Bechtel National*, NASA BCA No. 1186-7, 90-1 BCA ¶ 22,549 (1989) (internal citations omitted).

92 See Keane, P.J. and A.F. Caletka, *Delay Analysis in Construction Contracts,* Wiley-Blackwell, 2008, Section 3.4.4, pp. 102–104. Also, see Pickavance, Keith, *Delay and Disruption in Construction Contracts,* 3rd ed., T&F Informa (UK) Ltd, London, 2005, pp. 763–773, paragraph 19.11.

93 See *Pittman Construction Co.,* GSBCA Nos. 4897, 4923, 81-1 BCA ¶ 14,847 1980 WL 2643.

2.2.4 Preservation of Good Records

Calculating lost productivity damages that result from disruption or the cumulative impact of changes requires detailed records of how the contractor planned the work, and how the work was actually performed. However, not all contractors keep detailed cost records tied to specific work activities or groups of activities. Time cards may only identify that the worker performed pipefitting, but not where or what size pipe was installed. Also, the contractor's records may not identify the quantity of material installed on a daily or even weekly basis. The lack of detail in the contractor's cost records makes the application of the Measured Mile approach difficult to apply, thus leaving the analyst with no choice but to estimate the impact using less favored methods, such as industry studies, the Modified Total Cost approach, or the Total Cost approach.

Cause-effect analysis is difficult without adequate cost records. For example, absent actual man-hours and quantity installation measurement records, it will be difficult for the contractor to establish a relationship between a change or series of changes, events, actions, or inactions that allegedly were caused by the owner – or for which the owner had responsibility – and the loss of productivity that the contractor experienced.

Certain courts and arbitration tribunals have been forgiving of contractors that failed to keep adequate records, and have been willing to accept anecdotal evidence from witnesses involved in the project as proof of the cause of the alleged problems.[94] U.S. courts have been prepared to accept various methods for quantifying damages, and have tended to require strict proof of liability and causation by reference to records. For example, the VABCA found that the contemporaneous project records did not support the subcontractor's position:

> *The fact that proving the amount of productivity losses is recognized as being notoriously difficult does not abrogate PKC's fundamental responsibility to prove by a preponderance of the evidence that a Government action caused its labor to be less efficient than planned and the extent of that impact.... PKC primarily relies on the testimony of the PKC and USM project managers (Mr. MacClugage and Mr. Spors for PKC, Mr. Tammaro for USM) and its expert, Mr. Stynchcomb, to prove that a large portion of this overrun is due to VA-caused labor productivity losses. We consider the testimony of the project managers to be*

94 See Pickavance, Keith, *Delay and Disruption in Construction Contracts,* 3rd ed., T&F Informa (UK) Ltd, London, 2005, p. 659, paragraph 17.12, referencing *McAlpine Humberoak v. McDermott* (1992) 58 BLR 1 at pp. 15–20; *Ascon Contracting Ltd. v. Alfred McAlpine Construction Isle of Man Ltd* (1999) 66 Con LR 119; *Kvaerner Construction Ltd v. Egger (Barony) Ltd* (2000) QBD (TCC) Lawtel 8/9/2000(unreported); *Skanska Construction UK Ltd v. Egger (Barony) Ltd* (2002) EWHC 773 and (2004) EWHC 1748 (TCC); *Great Eastern Hotel Company Ltd v. John Laing Construction Ltd* (2005) EWHC 181 (TCC).

> *candid and forthright. However, PKC made little, if any, effort as we would ordinarily expect, to cite us to contemporaneous project records in support of the testimony. The Record in this appeal is one of the largest ever submitted to this Board. It contains all daily reports by the Government and Clark and its subcontractors, all CPM updates, voluminous correspondence, and all payment requests. Given this voluminous body of evidence, the parties were reminded of the importance of directing the Board's attention to the specific evidence that they believed supported their respective positions. The parties were instructed that it was their responsibility to provide proposed findings of all relevant facts specifically citing supporting Appeal File or Trial exhibits. PKC asserts here that the VA is liable for the loss of labor productivity resulting from it having to work in conditions where there was excessive water in the building and from the VA's failure to timely respond to RFIs. <u>Given the labor overrun that PKC knew had begun very early in the project, we find it difficult to believe that the contemporaneous documentation contained in the Record would not provide relevant evidence supporting both the fact that an impact on productivity occurred and the extent of that impact. Therefore, in this circumstance, we make the inference that the contemporaneous project records do not support PKC's position.</u>*[95] [underlining added]

The MCAA discusses separate cost coding of change orders and their impacts:

> *The contractor should determine whether it is practical to document the costs of a change using separate project job cost coding....When it is impractical to separately track costs associated with the change due to the effects on the base contract scope and/or cumulative impacts associated with multiple changes, it is helpful to document the impacts and costs through additional contemporaneous notes in the daily project logs,, timesheets, or other daily reports so that the portions of the work that are impacted can be more clearly identified in the absence of separate cost coding.*[96]

Although contractors sometimes try to use the Total Cost method or Modified Total Cost method to quantify its disruption and cumulative impact costs as a substitute for poor record-keeping, such tactics can be perilous. The contractor should maintain detailed quantity, man-hour, and cost records to demonstrate with particularity the elements of damages that it seeks to recover. These records are discussed in Chapter 6, Section 6.4 and Appendix C.

---

95 *Clark Construction Group Inc.*, 00-1 BCA ¶ 30,870, VABCA No. 5674 (April 2000), at p. 35 (Westlaw print).

96 "Change Orders Productivity Overtime – A Primer for the Construction Industry," MCAA, 2012, p. 5.

2.2.5 Overcome the Claim Premise Being Barred by Law or Contract

Depending on the jurisdictional law that governs the adjudication of unresolved claims, a claim for cumulative impact may be barred based on law, *i.e.*, such claims are not recognized by the law that governs the contract. This may be, for example, because the jurisdiction does not consider that cumulative impact claims are quantifiable with sufficient precision to be recoverable.[97] Legal representation is needed to attempt to overcome this challenge.

2.3 Arguments Against the Theory of Cumulative Impact

One author challenges the theory of cumulative impact claims and a contractor's entitlement to recovery on such claims,[98] in an article which has been cited by at least one court in its decision regarding a cumulative impact claim.[99] The author sets out a number of arguments, most notably that cumulative disruption does not exist:

> *Although disruption (i.e., the loss of productivity) is sometimes categorized as being either local or cumulative, there is no such thing as synergistic, greater-than-the-sum-of-its-parts cumulative disruption. At least insofar as entitlement is concerned, all disruption is "local." It is more accurate to refer to "unforeseeable" disruption rather than "cumulative" disruption. To use the term "cumulative" implies that the disruption is somehow causally different from local disruption and this is simply not the case. In fact, the typical cumulative disruption claim is nothing more than a "total productivity loss" claim and should be subject to the same prerequisites as "total cost" claims...*[100]

In support of his conclusions, the author provides several cogent arguments, some of which are summarized below:

- That the original justification used by U.S. Boards to award costs for cumulative impact – generally that impacts are difficult to quantify and cannot be recognized until project completion – is outdated and flawed. The contractor should be able to contemporaneously document the

---

97 For example, *"cumulative impact claims are too speculative to meet the requirements for possible recovery in Nevada and are not recognized under Nevada law."* (In re *Venetian Lien Litig.*, 2004 WL 326505 (Nev. Dist. Ct., 2004).

98 Finke, Michael R., "Claims for Construction Productivity Losses," 26 Pub. Cont. L.J. 311 (2008).

99 See *AMEC Civil, LLC v. DMJM Harris, Inc.*. Civil Action No. 06-64 (FLW), US District Court, D. New Jersey (June 2009).

100 Finke, Michael R., "Claims for Construction Productivity Losses," 26 Pub. Cont. L.J. 311 (2008), pp. 337–338.

changed working conditions and the actual productivity achieved under those conditions.[101]

- That the theory of cumulative disruption incorrectly considers the entire project to be a single "family" that can be disrupted by a given change. He argues that *"the existence of a shared working condition merely establishes potential causation,"* but the fact that *"...disruption was possible will not suffice to establish disruptive causation."*[102]

- That a change occurring late into a project cannot impact prior work or prior changes, yet to accept the concept of synergistic cumulative disruption, *"one must be willing to believe that a change order can travel back through time and affect past working conditions."*[103]

- That disruptive change orders may have "hidden costs" and be less than fully priced due to flaws in pricing, implementation, and management, but that these costs do not amount to synergistic cumulative disruption.[104]

- That many cumulative disruption claims are "total productivity loss" claims, and therefore must satisfy the four prerequisites for proving a "Total Cost" claim.[105]

This author's arguments and conclusions are useful for an owner defending a cumulative impact claim. He also proposes a format for disruption analyses and provides recommendations to enhance a contractor's position, both of which focus on properly demonstrating direct disruption.

2.4 Successful Claim Recovery

In *Bell BCI Company v. United States*,[106] the court granted the contractor the majority of its claimed costs for cumulative impact. The following are a few significant points from the court's decision:

---

101 *Id.* at pp. 325–326.

102 *Id.* at pp. 328–329.

103 *Id.* at pp. 329–330.

104 *Id.* at p. 331.

105 *Id.* at p. 333.

106 See *Bell BCI Co. v. United States*, No. 03-1613C, 81 Fed. Cl. 617, 2008 U.S. Claims LEXIS 116 (April 2008).

- The changes to the project were significant, both in terms of quantity and magnitude. There were 206 contract modifications issued, comprising over 730 Extra Work Orders.[107] The modifications added $21.4 million to the original contract price of $63.7 million, which represented a 33 percent increase.[108] Approximately nine months into construction, a new floor was added to the planned five-story building.[109]

- The court found that Bell had not expressly released its cumulative impact claim in any modification, and found that the owner's "accord and satisfaction" defense was without merit.[110]

- It was Bell's practice to track installed quantities, man-hours, and productivities at a cost account level for each week of the project.[111] Bell's quantum expert was able to use this detailed data to show Bell's earned and unearned man-hours by week, and align the results with the timing of the contract modifications. The expert's analysis showed, among other things, that Bell's productivity was near its as-bid productivity prior to substantial modifications, and that Bell's productivity degraded when major changes were introduced.[112]

- The court also accepted a Measured Mile analysis prepared by a Bell subcontractor, noting that it was an *"accepted method to prove a cumulative impact claim."*[113]

The successful outcome of Bell's cumulative impact claim resulted from its good project management practices during the project (reserving its rights and keeping detailed productivity records) and Bell's attorneys and its expert properly demonstrating causation and quantum.

---

107 *Id.* at p. 2.

108 *Id.* at p. 1 and 4.

109 *Id.* at pp. 1, 4, and 6.

110 *Id.* at p. 3.

111 *Id.* at p. 19.

112 *Id.* at pp. 19–20.

113 *Id.* at p. 35.

Chapter 3

Methods for Estimating Loss of Productivity Man-Hours for Disruption and Cumulative Impact Claims

Chapter 3

Methods for Estimating Loss of Productivity Man-Hours for Disruption and Cumulative Impact Claims

As previously discussed in Chapter 1, there are two distinct types of disruption: local (direct) and cumulative (indirect); the latter of which is commonly referred to as cumulative impact. Local disruption has been described as the direct, foreseeable impact of a specific change on the performance of the changed and/or the unchanged work. Cumulative impact, on the other hand, has been described as the indirect, unforeseeable impact of two or more changes or other impacting events on the performance of the work, and is exclusive of the local disruption that can be ascribed to a specific change.[114]

At the heart of a disruption and cumulative impact claim is the method or methods for estimating productivity loss. Most industry experts have concluded that using actual productivity data from the subject project is the best way to demonstrate a contractor's entitlement to a loss of productivity claim. Consequently, a Measured Mile analysis may be well suited to estimate the additional man-hours resulting from loss of productivity. In addition, many industry studies include research regarding the impact of overtime and changes on construction labor productivity, and several have further attempted to provide models to estimate such productivity losses. One group has researched the impact of changes on engineering man-hour productivity and construction labor productivity, and has provided separate models to estimate engineering productivity loss and construction labor productivity loss.

Productivity is a measurement of the rate of output per unit of time or effort, usually measured in either engineering man-hours or construction labor hours. Examples of productivity measurements are engineering drawings completed per man-hour or cubic yards of concrete placed per man-hour.

Production is the quantity of output (*i.e.*, units produced), whereas productivity is the measurement of the production. It is possible for a contractor to achieve 100 percent of its planned production, but not achieve its planned productivity. For example, a

---

114 See *Clark Concrete Contractors, Inc. v. General Services Administration*, 99-1 BCA ¶ 30280, GSBCA No. 14340 (1999), and *Centex Bateson Constr. Co.*, VABCA Nos. 4613, 5162, 5165, 99-1 BCA ¶ 30,153 (1998), aff'd, *Centex Bateson Constr. Co. v. West*, 250 F.3rd 761 (Fed Cir. 2000).

contractor could achieve its planned rate of production of 1,000 linear feet of large bore pipe per day in a process plant, but expend twice the amount of labor planned to accomplish this daily production rate. In this case, the contractor would be accomplishing 100 percent of planned production, but achieving only 50 percent of its planned productivity.

Productivity loss, therefore, occurs when a contractor is not accomplishing its planned rate of production (*i.e.*, the contractor is producing less than its planned output) per work hour of input. Thus, the contractor is expending more man-hours or crew hours per unit of production than originally planned.

In this Chapter 3, the following topics will be discussed:

- Causes of Productivity Loss (see Section 3.1);
- Contractor-Caused Problems that Contribute to Growth in Quantities, Man-Hours, and Costs (see Section 3.2);
- Industry Recommended Practices for Calculating Productivity Loss (see Section 3.3); and
- Methods for Estimating Productivity Loss (see Section 3.4).

3.1 Causes of Productivity Loss

AACE International has identified numerous potential causes of productivity loss, such as:[115]

- *Absenteeism and the missing man syndrome* – Absence of crew members may lower the overall crew productivity rate because of fewer resources or different skill levels of the remaining crew.
- *Acceleration (directed or constructive)* – Productivity can be negatively impacted by the extended use of overtime, additional shifts, or additional work crews that compromise effective management and coordination.

---

115 See "Estimating Lost Labor Productivity in Construction Claims," AACE International Recommended Practice No. 25R-03, April 13, 2004, pp. 4–6. Many of these same factors are identified by other industry publications, including "Change Orders Productivity Overtime: A Primer for the Construction Industry," MCAA, 2012, pp. 7, 77–78.

- *Adverse or unusually severe weather* – Productivity can be negatively impacted when work is delayed from good weather periods to bad weather periods, or when unusually severe weather is experienced.

- *Availability of skilled labor* – Productivity will likely be negatively impacted if skilled labor is unavailable and the contractor has to use less skilled labor than planned.

- *Changes, ripple impact, cumulative impact of multiple changes and rework* – Productivity is negatively impacted if the amount of change or rework is in excess of what is considered normal. Delays caused by changes, re-planning, and/or re-sequencing can disrupt unchanged work, which may not be considered in the direct costs of the change.

- *Competition for craft labor* – Productivity can be negatively impacted when there is unanticipated competition for the available skilled labor in the area of a project, causing skilled workers to leave a given project and the contractor to use less skilled replacement workers.

- *Craft turnover* – Learning curve productivity loss results from the continual turnover of labor.

- *Crowding of labor or stacking of trades* – Less than optimum work space may cause interference from other workers of the same trade or workers from other trades, resulting in productivity loss.

- *Defective engineering, engineering recycle and/or rework* – Ambiguous or erroneous drawings or specifications may delay work and cause productivity loss as crews wait for clear instructions or tear out and replace defective or incorrect work.

- *Dilution of supervision* – Splitting up crews to perform base work and changed work, or continual re-planning and re-sequencing of work, can result in less than effective supervision and lower productivity.

- *Excessive overtime* – Fatigue, increased absenteeism, decreased morale, reduced supervision effectiveness, and poor workmanship result from the excessive use of overtime, which causes productivity loss.

- *Failure to coordinate trade contractors, subcontractors and/or vendors* – Productivity loss may occur if subcontractors, materials, or equipment are not available when needed.

- *Fatigue* – Productivity declines as workers become tired, and more mistakes, injuries, and accidents than normal may occur.

- *Labor relations and labor management factors* – Productivity loss may result from union jurisdictional and industrial relations issues, safety issues, evacuation alarms, and late permit and access issues.

- *Learning curve* – Work stoppages and replacement of workers due to turnover or other problems causes labor to re-familiarize themselves with the work scope and location of tools and equipment, resulting in productivity loss.

- *Material, tools and equipment shortages* – Unavailability of material, tools, and equipment, or wrong or improperly sized tools and equipment, result in reduced productivity.

- *Overmanning* – More than planned labor personnel in an effort to recover from delays may result in less than effective supervision and reduced productivity.

- *Poor morale of craft labor* – Labor enthusiasm and productivity declines as work is continually changed and excessive rework is required.

- *Project management factors* – Inadequate scheduling and coordination of the work may result in shortage of labor and equipment when needed, resulting in productivity loss. Improper or poorly implemented procedures, poor site layout, or premature site mobilization may also result in productivity loss.

- *Out of sequence work* – By not planning the work in a logical manner, crews are moved around excessively and productivity declines.

- *Rework and errors* – Correction of work due to engineering errors or improperly performed installations results in productivity loss.

- *Schedule compression impacts on productivity* – Dilution of supervision, shortages of materials, tools or equipment to support the additional labor, increased difficulty in planning and coordinating the work, and shortages of skilled labor may result from attempts to perform the original scope of work in a shorter than planned period of time, or from performing additional work during the original schedule duration.

- *Site or work area access restrictions* – Productivity loss may occur if the work site is remote or has limited access.

- *Site conditions* – Productivity is negatively impacted by unsafe or difficult site conditions which hamper the efficient movement of equipment, materials, and labor around the site.

- *Untimely approvals or responses* – Failure of owners, general contractors, or design professionals to approve submittals or requests for information in a timely manner can delay the efficient execution of the work and reduce worker productivity.

Not all of the aforementioned causes of productivity loss are necessarily the responsibility of the owner and, therefore, depending on the terms and conditions of the contract, the additional man-hours incurred by the contractor as a result of certain causes of productivity loss may not be compensable. Also, certain of the aforementioned causes of productivity loss may be overlapping, *i.e.*, certain causes may be the result of other causes. For example, "Crowding of labor or stacking of trades" may be the result of "Schedule Compression Impacts," which may be the result of the contractor's attempt to recover time as a result of its "material, tools and equipment shortages." Therefore, it is important to isolate the potential multiple causes of productivity loss such that the compensable effect can be quantified.

Also, as stated by AACE International, the contractor should consider that the productivity loss inherent in performing change order work should be included as part of the costs of that change order:

> *Prior to initiating a loss of productivity analysis, the claimant should carefully consider whether the productivity loss can be recast as an impact of specifically definable extra work. If so, then such productivity loss ought to be incorporated into the estimate of the extra work and resolved in that manner.*[116]

3.2 Contractor-Caused Problems that Contribute to Growth in Quantities, Man-Hours, and Costs

The contractor is not entitled to recover costs resulting from problems for which it is responsible or, under most contracts, for costs resulting from "acts of god" or force majeure events. By performing a thorough analysis of the project documentation, and referencing the contractor's contractual responsibilities, an analyst should be able to determine actions or inactions by the contractor that may have led to increased direct costs, direct and indirect disruption and productivity loss, and delay costs.

---

116 "Estimating Lost Labor Productivity In Construction Claims," AACE International Recommended Practice No. 25R-03, April 13, 2004, p. 9.

Increases in the man-hours expended could have been caused, in part, by one or more of the following contractor-caused problems:

- Bid error, which when discovered later by the contractor caused an increase in the contractor's man-loading over its original plan, or caused re-sequencing of work;
- Contractor's failure to complete its design according to the specifications, resulting in rework;
- Late submission of construction plans, information required for permits, issued for construction drawings, specifications, and other contractor submittals, causing delay and potentially acceleration and productivity loss impacts;[117]
- Labor strikes caused by the contractor's improper labor practices;
- Idle time due to bad weather for which the contractor should have planned;
- Equipment down-time caused by the contractor's improper equipment operation or maintenance;
- Contractor's failure to expedite its vendors for timely delivery of materials and installed equipment;
- Inefficient management of the contractor's labor or that of its subcontractors;
- Contractor's failure to properly manage the time aspects of the project, including but not limited to the improper preparation of a logic-linked CPM schedule, failure to include all of the scope of work in the schedule, inadequate sequencing of the work, inadequate activity durations due to

[117] Owner review and response times to the contractor's submissions are often defined by the contract or its associated procedures. The time that an owner may reasonably need to review, provide comments, and ultimately approve the contractor's submittals is often a function of the quality and completeness of the contractor's submittal documents, and the contractor's adherence to a reasonable schedule for submittal of the documents requiring approval. Often, the owner requires a longer period to review the contractor's submittals because the contractor submitted its documents in large batches rather than over a reasonable period of time, as may be required by the contract. If this problem occurs, the contractor may be responsible for any resulting delay and productivity loss.

errors in the contractor's bid estimate for quantities and labor productivity, and improper statusing of the schedule;

- Rework, defective work, or unqualified craft labor which later caused an increase in man-loading over its original plan;
- The contractor's failure to supply the job with a sufficient number of workers earlier in the project because of the unavailability of qualified craft labor in the area, which as a result forced the contractor to increase its man-loading in later stages of the project to make up for its own labor problems; and
- Other contractor-caused delays which caused later increases in manpower loading over its original plan.

The existence of any one or more of these problems will influence the method that an analyst should utilize to calculate productivity loss man-hours and costs that are compensable to the contractor.

3.3 Industry Recommended Practices for Calculating Productivity Loss

AACE International concludes that arbitration panels are more favorably impressed by damage calculations related directly to the project in dispute and supported by contemporaneous project documentation.[118] Therefore, AACE International's recommended practice for preparing a lost productivity estimate is to utilize, if possible, one of the techniques listed in the category "Project Specific Studies," which include:

- Measured Mile Study;
- Earned Value Analysis;
- Work Sampling Method; and
- Craftsmen Questionnaire Sampling Method.[119]

---

118 For a more thorough discussion of this point, AACE International references Schwartzkopf, William and John J. McNamara, *Calculating Construction Damages,* 2nd Edition, Aspen Law & Business, New York, 2001, §1.03. See also Wickwire, Jon M., Thomas J. Driscoll, Stephen B. Hurlbut and Scott B. Hillman, *Construction Scheduling: Preparation, Liability and Claims,* 2nd Edition, Aspen Law & Business, New York, 2003, §12.04 et. seq.

119 "Estimating Lost Labor Productivity In Construction Claims," AACE International Recommended Practice No. 25R-03, April 13, 2004, pp. 11–12.

These methods are project specific and supported by personnel directly involved at the time of the dispute or when the disputed work was performed. Not all of these studies may be applicable to a cumulative impact claim analysis.

If there is insufficient information available from contemporaneous project documentation to support one of these techniques, the AACE International's recommended practice then is to use one of the methods listed under the category "Project Comparison Studies," which includes:

- Comparable Work Study; and
- Comparable Project Study.[120]

These methods, too, are project specific, but rely upon different forms of contemporaneous documentation. Also, not all of these studies may be applicable to a cumulative impact claim analysis.

Contemporaneous project documentation may not always be available to estimate lost productivity. In the absence of other proof of damages, the legal system that governs the project may allow estimates to be used to establish damages. However, even where the use of estimates is permitted, it is recognized that estimates are likely to be more readily subject to challenge than direct project costs. Of the damage estimating methods, AACE International's recommended practice is to first use one of the studies listed in the Specialty Industry Studies category. These are specialty studies of specific types of problems, and are generally based on some number of actual construction projects, as follows:

- Acceleration;
- Changes, Cumulative Impact and Rework;
- Learning Curve;
- Overtime and Shift Work;
- Project Characteristics;
- Project Management; and
- Weather.[121]

In Appendix B of its publication,[122] the AACE International lists a number of specialty studies relevant to the analysis of productivity loss associated with changes, cumulative impact, and rework. Of particular note are the following:

---

120 *Id.* at pp. 12–13.

121 *Id.* at pp. 13–14.

122 See "Estimating Lost Labor Productivity In Construction Claims," AACE International Recommended Practice No. 25R-03, April 13, 2004, Appendix B.

- Construction Industry Institute, "Quantitative Impacts of Project Change," CII SD-108, Austin, Texas, May 1995;

- Construction Industry Institute, "Quantifying the Cumulative Impact of Change Orders for Electrical and Mechanical Contractors," Research Summary 158-1, Austin, Texas, 2001;

- Hanna, Awad S., Jeffrey S. Russell, Erik V. Nordheim and Matthew J. Bruggink, "Impact of Change Orders on Labor Efficiency for Electrical Construction," Journal of Construction Engineering and Management, Vol. 125, No. 4, 1999;

- Hanna, Awad S., Jeffrey S. Russell, Timothy W. Gotzion and Erik V. Nordheim, "Impact of Change Orders on Labor Efficiency for Mechanical Construction," Journal of Construction Engineering and Management, Vol. 125, No. 3, 1999;

- Leonard, Charles A., "The Effects of Change Orders on Productivity," Concordia University, Montreal, Quebec, April 14, 1987; and

- Mechanical Contractors Association of America, "Change Orders, Overtime and Productivity – A Primer for the Construction Industry," Publication M3, Rockville, Md., 1968.[123]

AACE International cautions that, to utilize one of these studies, the causation of the lost productivity should be appropriate for the particular problem studied.[124]

If there is insufficient contemporaneous documentation to support a project specific study or a project comparison study, and further, if the loss of productivity stemmed from numerous, non-specific causes, AACE International's recommended practice is to employ a general industry study, which includes the following:[125]

- Mechanical Contractors Association of America (MCAA), "Labor Estimating Manual: Appendix B, Factors Affecting Productivity," Rockville, Md., August 1998;[126]

---

123 This document was updated by the MCAA in 2005, 2011, and again in 2012.

124 See "Estimating Lost Labor Productivity in Construction Claims," AACE International Recommended Practice No. 25R-03, April 13, 2004, p. 9.

125 *Id.* at pp. 14–15.

126 This document has been updated by the MCAA in 2005, 2011, and again in 2012 with "Change Orders Productivity Overtime – A Primer for the Construction Industry."

- National Electrical Contractors Association (NECA), "Manual of Labor Units," Bethesda, Md., 1976 and 2003;[127] and

- "Modification Impact Evaluation Guide," Department of the Army, Office of the Chief of Engineers, Washington, D.C., July, 1979. This manual addressed a number of factors the Corps was willing to discuss and negotiate when considering the forward pricing of change orders.[128]

3.4 Methods for Estimating Productivity Loss

This section discusses methods for estimating (approximating)[129] the loss of man-hours resulting from the impact of direct disruption, the cumulative impact of multiple changes, the impact of overtime, and the impact of overcrowding and congestion.

Methods for estimating the loss of man-hours resulting from the impact of direct disruption and cumulative impact are discussed in Section 3.4.1. The results from any of these methods may contain both direct disruption and cumulative impact man-hours. The Measured Mile method, for example, includes productivity comparisons between an un-impacted period and an impacted period of work, and the incremental loss of productivity may be attributable to direct disruption, cumulative impact, or both. Similarly, while formulaic approaches derived from project data such as the studies by Leonard or Ibbs are often associated with cumulative impact, the lost man-hours resulting from these studies may include direct disruption.[130]

---

127 This manual gave a job factor checklist addressing some 25 factors for consideration under certain circumstances. Current editions of the manual no longer contain this checklist, but a summary of the checklist can be found in Schwartzkopf, William, *Calculating Lost Labor Productivity in Construction Claims,* Wiley Law Publications, 1995, §11.3 at p. 128.

128 In endnote 53 of "Estimating Lost Labor Productivity In Construction Claims," AACE International Recommended Practice No. 25R-03, April 13, 2004, AACE International notes the following: *It should be noted that although the Corps of Engineers officially recognized this Guide as a valid means to assess claims for more than twenty years, on June 14, 1996 the Corps of Engineers officially rescinded the Modification Impact Evaluation Guide by issuance of Circular No. 25-1-244. At that time, the Corps claimed that the Guide "has been updated and is incorporated in other publications to include higher level regulations, training course materials and other command guidance."*

129 The loss of productivity man-hours must be based on estimates because the actual productivity loss from specific causes is usually not known.

130 See Section 3.4.1.3.7 for further discussion.

In Section 3.4.2, several industry studies related to overtime and overcrowding/congestion are addressed, which are two of the most common labor productivity impacts associated with acceleration.

3.4.1 Estimating Loss of Productivity Man-Hours Related to Direct Disruption and Cumulative Impact

In this Section, the following methods for estimating loss of productivity man-hour related to direct disruption and cumulative impact are discussed in detail.

- Measured Mile Analysis;
- Earned Value Analysis;
- Formulaic approaches which correlate the amount of project change to productivity loss, including the 1988 Leonard Study, the Construction Industry Institute studies, and the 2005 Ibbs Study;
- The use of disruption factors, including the MCAA Factors and the NAVSEA Guidelines; and
- System Dynamics Modeling.

3.4.1.1 Measured Mile Analysis

The Measured Mile analysis utilizes the contractor's actual performance on a portion of the work to determine what the full scope of work should have cost, absent any owner-caused impacts. While the method is still an estimate, its use of actual productivity data from the project lends credibility to the results.[131] The Measured Mile analysis is most effective when the comparison periods are close in time, involve very similar types of work, and occur in the same contract, so that like work is compared. For example, the court in *Lamb Engineering* stated:

> *As we understand it, a measured mile employs the rate of production achieved in a representative sample of actual operations to estimate the amount of effort which would have been required to perform work which cannot be directly measured. Its probative value necessarily depends upon the comparability of the circumstances surrounding the sample to the circumstances which would have prevailed for the work which could not be directly measured. In our opinion, its use is limited to circumstances where, as here, proof of actual costs of the work as contracted is not feasible.*[132]

---

131 See Nguyen, Long D. and William Ibbs, "Case Law and Variations in Cumulative Impact Productivity Claims," Journal of Construction Engineering and Management, ASCE, August 2010, p. 827.

132 *Lamb Engineering & Construction Company*, EBCA 97-2 BCA ¶ 29,207 at 145,348.

Most industry experts, arbitration panels, courts, and boards would agree that the Measured Mile method is the preferred method of proving and recovering loss of productivity costs.[133] The Measured Mile analysis is most appropriate for the analysis of construction work that requires a cyclical or repetitive working pattern, such as multi-story construction, pipeline construction, tunneling, road laying, and pile driving. However, detailed cost, man-hour, and installed quantity data is required to perform the estimation accurately.

If the appropriate data is available, the Measured Mile analysis is the best approach to pricing a loss of labor productivity claim resulting from direct disruption, as well as a loss of productivity claim arising from the cumulative impact of change orders. The MCAA also endorses this method for determining the loss of productivity.[134] The Measured Mile analysis is described in the MCAA's publication "How to Apply the Measured Mile method of Productivity Analysis."[135] The MCAA states: *"Simply put, this methodology computes inefficiencies by measuring a contractor's actual productivity rate achieved in a time frame or area of lesser impact and compares the contractor's actual productivity in a time frame or area of representative impact."*[136] Thus, the MCAA recognizes that the contractor should first collect the necessary data to perform a Measured Mile productivity analysis, and then apply it as the basis for the productivity loss estimation as a first resort.

The MCAA also states: *"accurate recording of a contractor's productivity is simply a management necessity to ensure profitability, irrespective of the portent of an inefficiency claim."*[137] However, the MCAA recognizes that such an analysis is not always possible. In addition, the MCAA recognizes that the Measured Mile method is an appropriate way to estimate

---

133 See, *e.g.,* Pickavance, Keith, *Delay and Disruption in Construction Contracts,* 3rd ed., T&F Informa (UK) Ltd, London, 2005, p. 578, paragraph 17.79; "The Society of Construction Law Delay and Disruption Protocol," Oxford, October 2002, p. 31; Keane, P.J. and A.F. Caletka, *Delay Analysis in Construction Contracts,* Wiley-Blackwell, 2008, Section 4.2.3, pp. 101, 104–106; "Change Orders Productivity Overtime – A Primer for the Construction Industry," MCAA, 2012, pp. 105–124; Schwartzkopf, William, *Calculating Lost Labor Productivity in Construction Claims,* John Wiley & Sons, Inc., New York, 1995, §2.09[A] and §10.4; "Estimating Lost Labor Productivity In Construction Claims," AACE International Recommended Practice No. 25R-03, April 13, 2004, p. 8.

134 See "Change Orders Productivity Overtime – A Primer for the Construction Industry," MCAA, 2012, pp. 105–124.

135 See "Change Orders Productivity Overtime – A Primer for the Construction Industry," MCAA, 2012, pp. 105–124.

136 "Change Orders Productivity Overtime – A Primer for the Construction Industry," MCAA, 2012, p. 74.

137 *Id.* at p. 79.

the impacts to labor productivity of changes to the unchanged work, *i.e.*, cumulative impact.[138]

The Measured Mile analysis is a comparison between a contractor's actual labor productivity during a relatively un-impacted period (the "Measured Mile") and its actual labor productivity during a period that is impacted by changes or other owner-caused events. The basis of the contractor's argument is that its decreased productivity during the impacted period is a result of owner-caused impacts, and the contractor is therefore entitled to the difference between its actual costs to perform the work and what it should have cost to perform the work, based upon its Measured Mile productivity.

The following steps should be employed by the analyst in preparing a Measured Mile analysis:

1. Determine the discipline (*e.g.*, pile driving, concrete work, or piping) and scope of work (*e.g.*, all piping, System 2100 piping, or six-inch carbon steel piping) to be used in the analysis.

2. Determine the Measured Mile period of performance where the work was not negatively impacted by problems that were caused by the owner.

3. Estimate the contractor's labor productivity in the Measured Mile period.

4. Estimate the contractor's "should have been" cost based upon actual quantities of work and its Measured Mile productivity during the impacted period, making any adjustments necessary for contractor-caused problems during the impacted period that were not present in the Measured Mile period.

5. Estimate the loss of productivity due to the alleged owner-caused impacts using the difference between the contractor's actual costs and its "should have been" costs, as adjusted for contractor-caused impacts.

6. Develop specific cause-effect matrices and narratives explaining how owner-caused impacts led to decreased labor productivity.

The disciplines chosen for the Measured Mile analysis must meet two common criteria: (i) they have been impacted by owner-caused problems, and (ii) the contractor

138 See "Change Orders Productivity Overtime – A Primer for the Construction Industry," MCAA, 2012, p. 89.

experienced a man-hour and cost overrun in completing the work for that discipline. The feasibility of the Measured Mile analysis is almost always driven by the detail in the contractor's job cost records and the total losses on the project. For example, when the contractor is experiencing loss of productivity impacts to its piping work, should the contractor prepare one Measured Mile analysis which encompasses all of piping, or multiple analyses for piping in each area of the project? As with the preparation of any claim, the more detail provided, the better the chance of recovery. However, the contractor must weigh the cost and time of preparing such analyses against potential recovery.

Steps 2 – 6 of the method described above will be explained by way of a hypothetical example. From the contractor's cost and schedule records, Table 3-1 was compiled, showing that the contractor overran its labor costs for Above-Ground Piping by $500,000:

Table 3-1: Above-Ground Piping Data

Description	Planned	Actual	Variance
Quantity (LF)	5,000	5,500	500
Man-Hours	10,000	20,000	10,000
Productivity (MH/LF)	2.0	3.64	1.64
Labor Cost ($)	$400,000	$900,000	$500,000
Duration (Weeks)	10	15	5

To determine the Measured Mile period, tables and graphics are used to detail the contractor's actual performance by time, measured by installed quantities, actual man-hours, and actual productivity. Figure 3-1 shows that the contractor nearly achieved its planned weekly installation quantities in Weeks 3 through 6 and Weeks 12 through 14. Based upon this graphic, both periods are candidates for the Measured Mile.

Figure 3-1: Cumulative Linear Feet Installed per Week – Above-Ground Piping

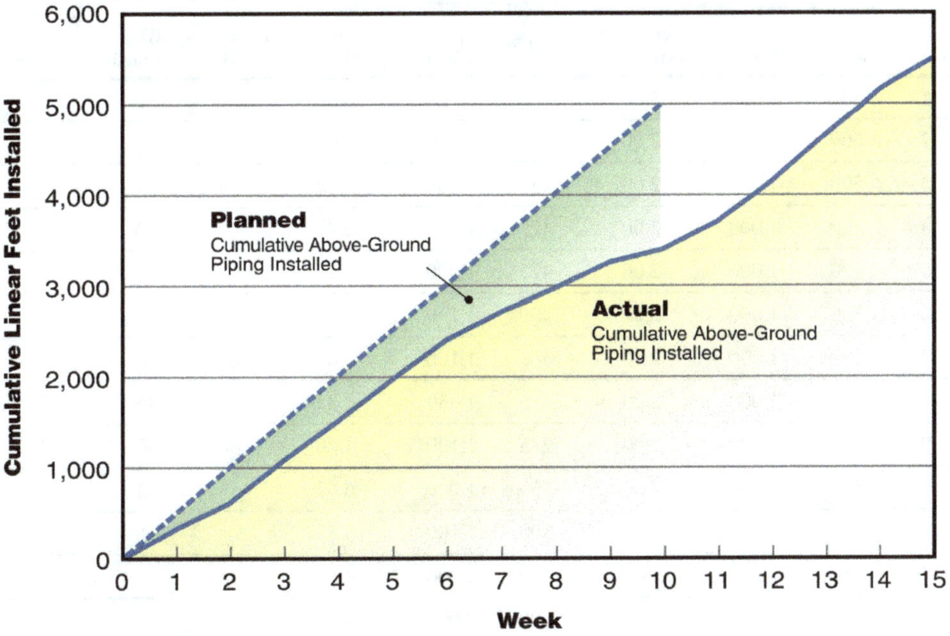

However, Figure 3-1 does not include the man-hours expended to install those quantities. In this example, the contractor accelerated during the later period (weeks 12–14) and, therefore, its productivity was far worse than planned. Table 3-2, showing the contractor's productivity by week, paints a much clearer picture of both the contractor's productivity loss and the events impacting its productivity.

Table 3-2: Planned and Actual Productivity – Above-Ground Piping

Week	Planned LF per Week	Planned MHs per Week	Planned Productivity per Week (MH/LF)	Actual LF per Week	Actual MHs per Week	Actual Productivity per Week (MH/LF)	Drawings Revised	RFIs Issued	Instances of Interference
1	500	1,000	2.00	300	900	3.00	–	2	–
2	500	1,000	2.00	300	900	3.00	–	–	–
3	500	1,000	2.00	450	1,000	2.22	5	1	–
4	500	1,000	2.00	425	1,050	2.47	–	3	–
5	500	1,000	2.00	475	1,000	2.11	7	1	–
6	500	1,000	2.00	450	1,100	2.44	–	–	–
7	500	1,000	2.00	300	1,100	3.67	67	10	–
8	500	1,000	2.00	275	1,050	3.82	125	15	7
9	500	1,000	2.00	275	1,100	4.00	80	17	4
10	500	1,000	2.00	150	1,000	6.67	55	11	8
11	–	–	–	300	1,600	5.33	–	8	3
12	–	–	–	450	2,000	4.44	15	6	1
13	–	–	–	500	2,000	4.00	–	4	–
14	–	–	–	500	2,400	4.80	–	–	–
15	–	–	–	350	1,800	5.14	–	–	–

From Table 3-2, four distinct periods of work are identified:

Weeks 1–2: Start-up Period. The contractor is unimpeded but slow to start due to mobilization of crews and learning curve. (Productivity[139]: 3.00 man-hours/LF)

Weeks 3–6: Likely Measured Mile Period. The contractor encounters some changed work, but no more than is typical for like projects, and achieves near its as-planned productivity of 2.0 man-hours per LF. (Productivity: 2.31 man-hours/LF)

Weeks 7–11: Impacted Period. The contractor encounters a multitude of changes, questions on installation due to poor design, and interference from other subcontractors, causing a reduction in productivity. (Productivity: 4.50 man-hours/LF)

139 Productivity is often measured in terms of units installed per man-hour of work, *e.g.*, 0.33 linear feet per man-hour. In this book, productivity is measured by man-hours per unit installed.

Weeks 12–15: Acceleration Period. Behind schedule, the contractor is forced to employ more men and overtime to catch up. While installed quantities per week are near plan, the man-hours are much greater than planned. (Productivity: 4.55 man-hours/LF)

Based upon productivities alone, the period from Weeks 3 through 6 is a reasonable Measured Mile period. However, the analyst must determine if this period is a representative sample of the entire scope of work. Three criticisms are likely:

1. The Measured Mile period chosen is not long enough, could be an anomaly, and/or was "cherry-picked" to show a productivity that could not be sustained.

2. The period reflects simplified work or work that was materially different from work outside the Measured Mile period.

3. The actual man-hours and costs in the analysis include work that was not impacted by the alleged owner-caused problems, some of which may not show a cost overrun in the contractor's cost report.

If, for example, in Weeks 3 through 6, the contractor installed mostly straight run piping, and subsequently installed mostly short run piping with many elbows and connections (which requires more man-hours to install per linear foot), Weeks 3 through 6 should not be used as the Measured Mile period to compare with later periods. In this instance, one would have to either choose a more detailed scope of work for the analysis, possibly by pipe size, or abandon the Measured Mile analysis to estimate damages.

In practice, these three likely criticisms are not easily overcome. Often the un-impacted period occurs early in the project, when the contractor's work is less complex and, therefore, no applicable comparison can be drawn between the un-impacted and impacted periods. Thus, choosing the correct scope of work to analyze can be an iterative process. If a supportable Measured Mile period cannot be determined for a given scope of work, one must analyze the components of the work to determine whether a more applicable comparison is possible.

Assuming that Weeks 3 through 6 in this example is a reasonable Measured Mile period, the Measured Mile productivity is estimated as 4,150 man-hours (the total man-hours in Weeks 3 – 6) divided by 1,800 linear feet (actual installed quantities in Weeks 3 – 6), or 2.31 man-hours per linear foot. This productivity, as well as the "should have been" linear footage per week and the "should have been" cost per week, are used to extrapolate the final "should have been" cost.

In this example, the records show that the work performed during Weeks 1 through 6 was not impacted by owner-caused problems and, therefore, the "should have been" costs are equal to the contractor's actual costs. For all remaining weeks, the Measured Mile productivity of 2.31 man-hours per linear foot, as well as the "should have been" progress per week of 450 linear feet[140] and the "should have been" man-hours expended per week of 1,038,[141] are used to estimate the final cost. Table 3-3 provides the final estimate of the "should have been" cost.

Table 3-3: "Should Have Been" Costs – Above-Ground Piping Labor

Week	Should Have Been LF per Week	Should Have Been Cumulative LF	Should Have Been MHs per Week	Should Have Been Productivity per Week (MHs/LF)	Should Have Been Costs per Week*	Source
1	300	300	900	3.00	$36,000	Actual Data
2	300	600	900	3.00	$36,000	Actual Data
3	450	1,050	1,000	2.22	$40,000	Actual Data
4	425	1,475	1,050	2.47	$42,000	Actual Data
5	475	1,950	1,000	2.11	$40,000	Actual Data
6	450	2,400	1,100	2.44	$44,000	Actual Data
7	450	2,850	1,038	2.31	$41,500	Measured Mile Data
8	450	3,300	1,038	2.31	$41,500	Measured Mile Data
9	450	3,750	1,038	2.31	$41,500	Measured Mile Data
10	450	4,200	1,038	2.31	$41,500	Measured Mile Data
11	450	4,650	1,038	2.31	$41,500	Measured Mile Data
12	450	5,100	1,038	2.31	$41,500	Measured Mile Data
13	400	5,500	922	2.31	$36,889	Measured Mile Data
Totals	**5,500**	–	**13,097**	**2.38**	**$523,889**	

* Based on a $40/hour base labor rate (no overtime premiums)

Note that this "should have been" cost of $523,889 assumes that there were no contractor-caused problems or inefficiencies in Weeks 7 through 13 other than those same contractor-caused problems or inefficiencies that were included in, and affected productivity during, the Measured Mile period. If the contractor-caused problems that occurred during Weeks 7 through 13 were similar to those that occurred during

140 The average installed quantities in the four-week Measured Mile period, or 1,800 lf/ 4 weeks = 450 lf per week.

141 The average man-hour expenditure in the four-week Measured Mile period, or 4,150 mhs/4 weeks = 1,037.5 mhs per week.

the Measured Mile period, the cost of the problems would be included in the "should have been" cost. However, if these contractor-caused problems during Weeks 7 through 13 were significant, and persisted throughout the period, the "should have been" cost as estimated would be unreasonable, as it would not include the significant costs of the contractor-caused problems. In this example, assume that no significant contractor-caused problems occurred during Weeks 7 through 13.

Therefore, the contractor's compensable damages claim for Above-Ground Piping would be $376,111 ($900,000 actual cost less $523,889 "should have been" cost), less any amount paid by the owner for any loss of productivity, overtime, or other costs that were previously paid to the contractor by the owner via approved change orders, such as a change order for the cost of the additional 500 feet of piping. The contractor's non-compensable cost would be $123,889, or the difference between its planned costs and its "should have been" costs, and would likely be attributed to bid error or contractor-caused performance problems. This non-compensable cost is justified because, even during the un-impacted period, the contractor did not meet its planned productivity.

The final step is to show that the owner-caused impacts – in this case design changes, interference by other contractors, and acceleration – actually caused the contractor's loss of productivity. A good start is to align the contractor's weekly productivity with the weekly impacts, as shown in Table 3-2 above, but this step will be further reviewed in Chapter 6.

If an un-impacted period of time is not available to use as the Measured Mile period, baseline productivity may be established by selecting a set of data points that are statistically most likely to represent undisrupted progress. In addition, an analysis may utilize productivity data from one or more other projects that were performed by the same contractor to establish the baseline, un-impacted productivity where it can be shown that the work was very similar to the work in the subject contract; the working conditions, equipment, and methods employed were the same; and the labor crew and supervision were of similar experience.

One advantage of the Measured Mile method is that it encompasses the impacts of all labor inefficiencies. In the above example, if the Measured Mile method could not be used, the analyst might have to estimate separately the labor inefficiencies caused by cumulative impact of change orders, the effects of overtime on productivity, and the impact of interference. These separate analyses would offer a less reliable result than that derived by using the Measured Mile method. However, if the contractor did not reserve its rights for recovery of costs that were caused by any one of the many impacts that could have occurred during the analysis period, use of the Measured Mile, even if properly prepared, may not lead to recovery of additional costs.

The Measured Mile method should be the preferred choice for a contractor submitting a loss of productivity claim, including direct disruption and/or cumulative impact, if it has collected the necessary data to support such an analysis. The contractor's entitlement to this claim may be dependent on whether it has provided notice and properly reserved its rights. In many instances, however, the Measured Mile method cannot be used, typically for three reasons: (i) there exists no un-impacted period of work to establish a baseline; (ii) the contractor's cost, man-hour, and installed quantity data lacks sufficient detail to determine productivities by time and relevant scope of work; and (iii) there are contractor-caused performance problems in the impacted period, and they cannot be distinguished from the owner-caused impacts.

While other methods, some of which are discussed herein, can be utilized to calculate loss of productivity, the Measured Mile method also requires detail from the contractor's cost records. A contractor seeking to recover its loss of productivity costs using the Measured Mile method must keep adequate cost, man-hour, and installed quantity data, at least to the level of detail shown in the example above.

3.4.1.2 Earned Value Analysis

Productivity loss can also be measured using an Earned Value analysis, which can be effective if productivity data is not available for a Measured Mile analysis. Unlike the Measured Mile method, in which the contractor's actual performance (during a specific period) is used as the basis for comparison, the Earned Value analysis relies upon comparisons with the contractor's plan or budget. The inefficient man-hours are calculated as the difference between the actual man-hours expended and the actual man-hours earned (based on the plan or budget) for the period of the impact.[142]

Periodic Earned Value performance and progress reports typically use installed quantity or weighting factors and progress information for individual accounts or commodities to calculate earned man-hours, which are summarized and compared to total budget estimate man-hours to calculate composite project progress. In addition, earned man-hours are often compared to actual hours at the detailed account, discipline, and total project levels to measure actual labor performance against the budget. If budget estimate information has been associated with schedule activities, work hour accomplishment based upon schedule activity progress can be summarized and compared to earned budget hours to determine if the project is behind or ahead of schedule, and to what extent. This report is valuable as a tool for the contemporaneous evaluation of progress and labor

---

142 See Nguyen, Long D. and Ibbs, William, "Case Law and Variations in Cumulative Impact Productivity Claims," Journal of Construction Engineering and Management, ASCE, August 2010, p. 830.

performance, and can be used for damage analyses related to man-hour growth and productivity losses for individual accounts.

Periodic productivity performance reports are used to compare planned budget estimate unit labor productivity for various material commodities and deliverables to actual unit labor productivity, usually expressed as man-hours per unit of commodity or deliverable. This report is used contemporaneously to identify commodities and deliverables that are experiencing poorer than planned productivity. It is also often used for claim quantum calculations associated with disruption- and productivity-related man-hour and cost growth for cost accounts that include the subject commodities or deliverables.

In the sample project shown in Table 3-2 above, the contractor experienced significant disruption to its work in Weeks 7 through 15. During this period, the contractor expended 14,050 actual man-hours to install 3,100 lineal feet of piping (a productivity of 4.53 man-hours per lineal foot). During this same period, using its planned productivity rate of 2.0 man-hours per lineal foot, the contractor earned 6,200 man-hours (3,100 lf x 2.0 man-hours/lf). Thus, the contractor experienced productivity losses of 7,850 man-hours (14,050 actual man-hours less 6,200 earned man-hours).

The contractor's claim amount would have to include the overtime premiums paid for the acceleration period in Weeks 12 – 15, and in this example would total $412,350, as shown in the Table 3-4 below.

Table 3-4: Claim using Earned Value Analysis for Above-Ground Piping Labor

Description	Man-Hours	Average Labor Rate	Labor Costs
Actual	14,050	$47	$660,350
Earned	6,200	$40	$248,000
Claimed	**7,850**		**$412,350**

The Earned Value analysis, in this example, yields a claim value that is greater than the value quantified using the Measured Mile analysis ($376,111). This is the expected result, given that the Measured Mile productivity (2.31 man-hours/lf) was worse than the planned productivity (2.0 man-hours/lf).

As with any quantum method that utilizes the contractor's plan or budget data, the planned productivity used in the Earned Value analysis must be proven to be reasonable and achievable. Furthermore, as with the Measured Mile method, the Earned Value analysis is most accurate and persuasive when applied to discrete scopes of work over discrete periods of impact.

3.4.1.3 Industry Studies on the Impact of Changes (Formulaic Approaches)

The construction industry has produced many studies attempting to quantify the impact of changes to productivity. If a contractor's records are insufficiently detailed such that Measured Mile or Earned Value methods of estimating productivity loss cannot be utilized, the contractor can employ industry studies to estimate the cumulative impact of changes. Examples of these industry studies include, but are not limited to, the following:

- Leonard, Charles A., "The Effects of Change Orders on Productivity," M.S. Thesis, Concordia University, Montreal, Quebec, 1988 and related article: Moselki, Leonard & Fazio, "Impact of Change Orders on Construction Productivity," Canadian Journal of Civil Engineering, Volume 18, 1991;

- Construction Industry Institute, "The Impact of Changes on Construction Cost and Schedule," Publication 6-10, April 1990;

- Hester, Weston T., John A. Kuprenas, and T.C. Chang, "Construction Changes and Change Orders: Their Magnitude and Impact," Construction Industry Institute, SD66, October 1991;

- Project Change Management Research Team, "Quantitative Effects of Project Change," CII Publication 43-2, May 1995;

- Ibbs, C.W. & Walter E. Allen, "Quantitative Impacts of Project Change," CII Source Document 108, May 1995;

- Ibbs, William C., "Quantitative Impacts of Construction Change: Size Issues," ASCE Journal of Construction Engineering and Management, 123, 3, 1997;

- Hanna, Awad S., Jeffrey S. Russell, Timothy W. Gotzion, and Erik V. Nordheim, "Impact of Change Orders on Labor Efficiency for Mechanical Construction," Journal of Construction Engineering and Management, ASCE 125: 3 (1999b);

- Hanna, Awad S. Jeffrey S. Russell, Erik V. Nordheim, and Matthew J. Bruggink, "Impact of Change Orders on Labor Efficiency for Electrical Construction," Journal of Construction Engineering and Management, ASCE 125: 4 (1999b);

- Hanna, Awad S., "Quantifying the Cumulative Impact of Change Orders for Electrical and Mechanical Contractors," CII Research Report 158-11, February 2001;

- Ibbs, William, "Impact of Change's Timing on Labor Productivity," Journal of Construction Engineering and Management, ASCE 131:11, 2005a;

- Ibbs, William, Long D. Nguyen, and Seulkee Lee, "Quantified Impacts of Project Change," Journal of Professional Issues in Engineering Education and Practice, ASCE, January 2007; and

- Ibbs, William and Gerald McEniry, CFCC PSP, "Evaluating the Cumulative Impact of Changes on Labor Productivity – an Evolving Discussion," Cost Engineering, AACE International, Vol. 50/No. 12, December 2008.

In this Section, the following methods for estimating the productivity loss related to cumulative impact are discussed in detail:

- 1988 Leonard Study (see Section 3.4.1.3.1);

- 1990 CII Study – CII Publication 6-10 (see Section 3.4.1.3.2);

- 1995 CII Study – CII Publication 43-2 and CII Source Document 108 (see Section 3.4.1.3.3);

- 2001 CII Study – CII Research Summary 158-1 and CII Research Report 158-11 with reference to Hanna's 1999 papers (see Section 3.4.1.3.4); and

- 2005 Ibbs Study and related 2007 and 2008 papers (see Section 3.4.1.3.5)

Courts have referred to these studies as formulaic approaches, because the studies provide formulas wherein one can estimate productivity loss by inputting only a few variables, such as the quantity of change order man-hours and the actual man-hours expended. Section 3.4.1.3.6 provides discussion of the general view of formulaic approaches by the courts. The uncertainties in using these studies and their formulaic approaches are discussed in Section 3.4.1.3.7. Chapter 4 provides example calculations using these studies.

3.4.1.3.1 1988 Leonard Study

A study entitled "The Effects of Change Orders on Productivity," sometimes referred to as the Leonard Study, was prepared in 1988 as a master's thesis by Charles A.

Leonard at Concordia University, Quebec, Canada. Leonard analyzed data from 90 cases drawn from 57 different construction projects, mostly industrial or building projects,[143] to identify and quantify the effects of change orders on productivity.[144] A paper co-written by Leonard in 1990 expanded upon the conclusions drawn from his thesis.[145]

Leonard attempted to correlate three separate relationships between change orders and loss of productivity: (i) frequency of change orders; (ii) average change order value; and (iii) the percentage of change order man-hours compared with base scope man-hours.[146] Only the third relationship could be proven statistically relevant:

> *The results indicate a significant direct correlation between the labor component of change orders and the loss of productivity, for both civil/architectural and electrical/mechanical works. These losses are exacerbated by the added presence of other major causes of productivity losses such as acceleration and inadequate scheduling and coordination.*[147]

Leonard prepared three models to predict the loss of productivity on other projects: one model for civil/architectural work, one for electrical/mechanical work, and one for a combination of the two. The models show that a direct (straight-line) relationship exists between change order man-hours and loss of productivity. The study notes: *"In appropriate cases experiencing greater than 10% to 15% in change orders, these models can be used to estimate productivity losses of labour at the micro level."*[148] Leonard's data is based only on additive changes.[149]

Leonard's findings show that electrical and mechanical work are more negatively impacted than civil and architectural work as change order man-hours increase.

---

143 Leonard states, at p. 11, *"excluded from the present study are the following: one-of-a kind, unusually complex and highly specialized industrial projects such as nuclear power plants; offshore exploration structures; megaprojects of any type; and heavy civil engineering-construction projects such as highways, tunnels, bridges, rail and marine ports."*

144 See Leonard, Charles A., "The Effects of Change Orders on Productivity," M.S. Thesis at Concordia University, 1988.

145 See Moselki, Leonard & Fazio, "Impact of Change Orders on Construction Productivity," Canadian Journal of Civil Engineering, Volume 18, 1991.

146 See Leonard, Charles A., "The Effects of Change Orders on Productivity," M.S. Thesis at Concordia University, 1988, p. 89.

147 Moselki, Leonard & Fazio, "Impact of Change Orders on Construction Productivity," Canadian Journal of Civil Engineering, Volume 18, 1991, p. 1.

148 Leonard, Charles A., "The Effects of Change Orders on Productivity," M.S. Thesis at Concordia University, 1988, p. iv.

149 See McEniry, Gerald, "The Cumulative Effect of Change Orders on Labour Productivity – The Leonard Study 'Reloaded'," The Revay Report, Volume 26, No. 1, May 2007, p. 6.

Leonard notes that this difference is attributable to *"the level of skill required to perform the work, complexity of the work, and the interdependency of the work activities."*[150]

Leonard's model for electrical/mechanical work is shown in Figure 3-2.[151] Leonard's model for civil/architectural work is shown in Figure 3-3.

Figure 3-2: Leonard Study – Effects of Change Orders on Productivity Electrical/Mechanical Work

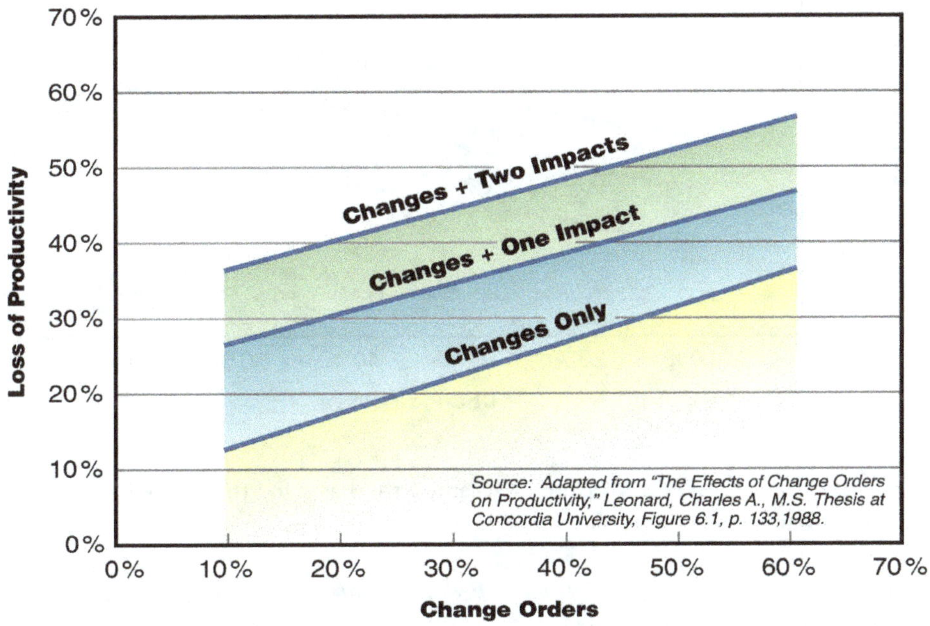

150 Moselki, Leonard & Fazio, "Impact of Change Orders on Construction Productivity," Canadian Journal of Civil Engineering, Volume 18, 1991, p. 6.

151 See Leonard, Charles A., "The Effects of Change Orders on Productivity," M.S. Thesis at Concordia University, 1988, p. 133.

Figure 3-3: Leonard Study – Effects of Change Orders on Productivity Civil/Architectural Work

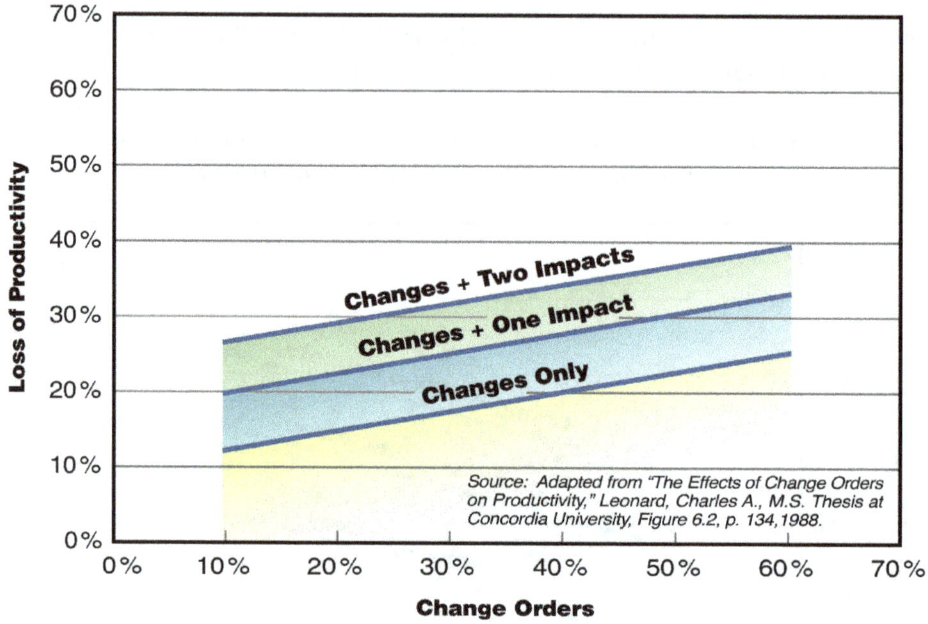

One commentator, Gerald McEniry, explained the basis for the various relationships in Leonard's curves.

> *Leonard measured the "percent change orders" (represented by the x-axis) in terms of the sum of change order hours divided by "actual" hours spent on the original contract (i.e., unchanged) work. Actual contract hours were not the estimated base contract hours, but rather the total actual hours less change order hours less any hours attributable to contractor's own inefficiencies. The actual hours, therefore, inherently included the impact of the change orders on the original contract (i.e. unchanged) work...*[152]

McEniry further explained how Leonard determined the productivity loss from the project data:

> *Leonard calculated the loss of productivity using the differential method of calculation ("measured mile") whenever possible. He first determined the "normal" hours from which the lost hours could then be established. In cases where the normal hours could not be determined, "earned" hours (which could be equated to normal hours) were used to calculate the lost hours. If the contractor's estimate was considered reasonable, then earned hours were*

152 McEniry, Gerald, "The Cumulative Effect of Change Orders on Labour Productivity – The Leonard Study 'Reloaded'," The Revay Report, Volume 26, No. 1, May 2007, p. 2.

> *the estimated (budgeted) hours. If the contractor's estimate was not reasonable, the earned hours were modified to put them in line with other bids.*
>
> *Leonard then calculated the productivity index ("PI") as the ratio of earned hours vs. actual base contract hours. Loss of productivity ("LOP") was calculated as follows:*
>
> *LOP = (1 – PI) x 100%*
>
> *Loss of productivity was also the ratio of the unproductive labour hours to the actual labour hours spent on the original contract (i.e. unchanged) work. Loss of productivity and the productivity index were illustrated on the y-axis.*[153]

McEniry then provided an example of how the Leonard Study curves are used to estimate productivity loss man-hours:

> *For example, consider a case where the normal or earned number of hours required for a project was determined to be 10,000 hours but that 20,000 hours were actually spent, including 6,000 hours worked and paid on change orders. Assuming for simplicity that no inefficiency hours could be directly attributed to the contractor alone, removing the change order hours leaves 14,000 hours actually worked on the base contract (including any loss of productivity due to the changes). The "percent change orders" is 43% (6,000 change order hours ÷ 14,000 actual base contract hours). The productivity index is 0.71 (10,000 earned hours ÷ 14,000 actual base contract hours) and the loss of productivity is 29%. As such, 4,000 hours were lost productivity out of a total of 14,000 hours actually worked on the base contract.*[154]

As shown in Figure 3-2 and Figure 3-3, Leonard's analysis also provided differing levels of productivity loss projections based on "changes only," "changes plus one impact," and "changes plus two impacts." Leonard identified the "major causes" as change orders; inadequate scheduling and coordination by owners, general contractors, and construction managers; acceleration; changes in sequence; late supply of information, equipment, or material; increased complexity of the work; and the ripple effect of change orders issued to other contractors.[155]

McEniry concludes that *"some of these so called 'major causes' may often be the impact of the changes themselves and that the distinction of results into separate levels of impact is not a simple task."*[156] He also noted that *"this distinctive approach, with three significantly different trend*

153 *Id.*

154 *Id.*

155 See Leonard, Charles A., "The Effects of Change Orders on Productivity," M.S. Thesis at Concordia University, 1988, p. 121.

156 McEniry, Gerald, "The Cumulative Effect of Change Orders on Labour Productivity – The Leonard Study 'Reloaded'," The Revay Report, Volume 26, No. 1, May 2007, p. 2.

lines, was not generally adopted by other researchers who have preferred to group all data together and attempt to explain the scatter in the data as the result of other factors."[157] Thus, before asserting that there may be multiple additional causes of productivity loss along with the effect of changes, the contractor will need to establish a supportable cause-effect analysis, as discussed in Chapter 7.

Leonard's Study has been criticized for several reasons, as follows:[158]

- The research was developed using projects that had reached the dispute stage, which may introduce a bias toward projects suffering important productivity losses;
- Some consider that the size and number of the projects studied were too small and insufficient to extrapolate to the industry as a whole;[159]
- The study cannot be used to predict loss of productivity when change order hours are less than 10 percent or greater than 60 percent of the original contract hours;[160]
- The study does not address the issue of the timing of the changes; and
- Other factors, such as the weather, labor conditions, and code requirements, are particular to the Canadian industry and not necessarily representative of the American industry.

Courts and boards reject contractor's quantum calculations for either using the wrong method, or using the correct method improperly. While Ibbs and likely others have successfully used Leonard,[161] courts and boards often reject Leonard as the basis for proving damages. In one case, the board appointed a Special Master, who dismissed the study as *"wholly unreliable because it is based on unfounded assumptions, biased and incomplete*

157 *Id.* at pp. 2–3.

158 See McEniry, Gerald, "The Cumulative Effect of Change Orders on Labour Productivity – The Leonard Study 'Reloaded'," The Revay Report, Volume 26, No. 1, May 2007, p. 3.

159 The data set only included 57 projects with an average size of $4 million, and all projects were in Canada.

160 The range of results of the study is 10 percent to 60 percent change orders.

161 See Ibbs, William, "The Cumulative Impact of Change on Construction Labor Productivity," Construction Claims Advisor, March 2, 2008, p. 4. Ibbs stated: "...the author of this present article has successfully used both the Leonard and Ibbs data presented in this column in trial, arbitration, and mediation settings."

data, and speculative conclusions."[162] In *J. A. Jones Construction*, the contractor's initial claim for the cumulative impact of changes was calculated based upon the Leonard Study. The board commented that the Leonard Study was "*based upon much smaller projects, of different scope, and trades than the project in question,*" and that "*no court has adopted the Leonard Study approach in measuring productivity loss/inefficiency.*"[163]

McEniry points out that, in a September 2006 article,[164] authors Harmon and Cole disagreed with some of these criticisms. However, McEniry states that no published court decisions in the U.S. or Canada indicate that the Leonard approach has been accepted in predicting the cumulative loss of productivity. This article also refers to the aforementioned published opinion discussing the Leonard Study (*Appeals of J. A. Jones Const. Co*),[165] where the method was rejected as being inapplicable to heavy civil engineering construction projects, and it was suggested that the expert testifying about the study was apparently unfamiliar with its details and limitations, which ultimately led to it being abandoned.

One author reports on a study in which the projected productivity loss that was estimated from using the Leonard Study was validated.[166] The author projected the impact of changes using the Leonard method in a forward pricing exercise. At project completion, the actual productivity loss, based on actual hours expended, was determined to be within 2 percent of the projection.

If the analyst is unfamiliar with the details of the study and the limitations to its use, the Leonard Study can be misused. These potential application problems include the following:[167]

- No effort is made by the analyst to use the differential or Measured Mile method where possible, contrary to Leonard's recommendation;

---

162 *Aetna Casualty & Surety Co. v. George Hyman Construction Co., U.S. Dist.*, LEXIS 22627 (E.D. Pa. 1998).

163 *J. A. Jones Construction Co.*, 00-2 BCA ¶ 31,000, ENGBCA Nos. 6390-1, 6386-7, 6348, 6388-9, 2000 WL 1014011 at pp. 46–47.

164 See Harmon, Dr. K.M.J. and B. Cole, "Loss of Productivity Studies – Current Uses and Misuses: parts 1 and 2," Construction Briefings, August and September 2006.

165 See *J. A. Jones Construction Co.*, 00-2 BCA ¶ 31,000, ENGBCA Nos. 6390-1, 6386-7, 6348, 6388-9, 2000 WL 1014011.

166 See Schwartzkopf, William, *Calculating Lost Labor Productivity in Construction Claims*, Second Edition, Aspen Publishers, 2004, Chapter 4.06, The Effect of Change Orders of Productivity, p. 94.

167 See McEniry, Gerald, "The Cumulative Effect of Change Orders on Labour Productivity – The Leonard Study 'Reloaded'," The Revay Report, Volume 26, No. 1, May 2007, p. 3.

- Assumptions by the analyst that all overruns in hours are the result of owner-caused changes may be inappropriate, and no adjustments are made for other factors such as estimating errors or contractor-caused performance problems. This problem would also apply to most other methods to estimate productivity loss;

- Miscalculation by the analyst of "percent change orders" as either: (i) a function of the dollar value of changes instead of hours, or (ii) a portion of estimated hours rather than actual base contract hours; and

- Failure by the analyst to demonstrate a cause-effect link between the alleged inefficiency and the change. Other factors not related to owner directed changes (weather, project conditions, union problems, delays) could be responsible for the lost productivity on a construction project.

While it remains a commonly cited industry study regarding cumulative impact of changes, courts and boards often reject Leonard as the basis for proving damages. Therefore, the Leonard Study models may be used by a contractor to negotiate equitable adjustments, or at best, to corroborate the results of other damages methods, but it is not advised that the Leonard Study be used as the sole basis for calculating a contractor's loss of productivity cumulative impact claim.

3.4.1.3.2 1990 CII Study

A CII study from 1990 took a different approach to investigating loss of productivity. Instead of compiling and comparing data from dozens of different projects, the study focused on measuring the impact of interruption on a pipe insulation crew completing a very discrete scope of work. The study concluded that *"productivity drops rapidly with increased frequency of interruptions."*[168] For the detailed tasks analyzed, the study found that one interruption reduced productivity by about 9 percent, two interruptions by around 39 percent, and three interruptions by 69 percent, not including the actual interruption time.[169]

The study went further to examine the "ripple effect" of change orders, both on concurrent work and follow-on work. While the study was unable to quantify this "ripple," the observations are noteworthy. Regarding the impact to sequential succeeding activities, it noted:

---

168 Construction Industry Institute, "The Impact of Changes on Construction Cost and Schedule," Publication 6-10, April 1990, p. 9.

169 See Construction Industry Institute, "The Impact of Changes on Construction Cost and Schedule," Publication 6-10, April 1990, p. 9.

> *Research conducted as part of Study 2…involved the sequential accounts of concrete, structural steel, and above-ground piping. In analyzing the data from these observations, it was noted that a loss of productivity in an account due to a change was followed almost immediately by a loss of productivity in a succeeding account.*[170]

Regarding the impact on adjacent concurrent work, the study explained:

> *Lowered productivity seems to be contagious. Just as an accident stops all work in an area, any type of interruption in one activity can affect surrounding activities. Crews affected by changes will not keep their dissatisfaction to themselves and may interrupt adjacent crews. Interferences with adjoining crews also may be created by the change.*[171]

While this study is unlikely to help the contractor quantify its cumulative impact claim, the conclusions drawn can be used to enhance the cause-effect proof needed to support such a claim.[172] Interruptions in one form or another, and their impact on other work, are at the very heart of a cumulative impact claim.

3.4.1.3.3 1995 CII Study

In 1995, the Construction Industry Institute (CII) published a study that was performed by a research team at the University of California at Berkeley.[173] The research team performed a statistical analysis of 104 industrial projects from 35 different companies (15 contractors and 20 owners) valued from US$3.2 million to US$1.2 billion in an attempt to quantify the impact of project change. The median value of the projects was US$44 million.[174] The projects used in this study were on average much larger than those projects used by Leonard. Further, this study analyzed the impacts of engineering and construction change orders separately, and is the only cumulative impact study known by the authors that provides separate productivity loss data for engineering. The project database included both disputed and undisputed projects.[175]

---

170 *Id.* at p. 15.

171 *Id.*

172 See Chapter 7 herein.

173 See Ibbs, C.W. and Walter E. Allen, "Quantitative Impacts of Project Change," CII Source Document 108, May 1995.

174 See McEniry, Gerald, "The Cumulative Effect of Change Orders on Labour Productivity – The Leonard Study 'Reloaded'," The Revay Report, Volume 26, No. 1, May 2007, p. 3.

175 See Ibbs, William, "Impact of Change's Timing on Labor Productivity," Journal of Construction Engineering and Management, ASCE 131:11, 2005a, p. 1220.

In a supplemental paper, CII stated that *"the findings of the study are helpful in benchmarking the amount of change and the expected productivity on a given project against the studied projects."*[176]

In this 1995 study, CII provided the following definitions:

- *Productivity = the work-hour inputs per unit of measure (e.g., work-hours expended per cubic yard of concrete installed)*
- *Productivity Index (PI) = earned work-hours/ expended work-hours*
- *The "percent change" is defined as the number of work-hours expended on "authorized changes that originated during the construction phase" divided by total number of work-hours expended for construction*[177]

The study includes several models that show a direct relationship between change order man-hours and loss of productivity. Of 16 tests conducted, the study found that the following seven relationships were statistically significant and verifiable: Construction Change vs. Construction Productivity; Engineering Change vs. Engineering Productivity; Construction Change vs. Total Productivity; Engineering Change vs. Total Productivity; Total Change vs. Total Productivity; Total Change vs. Construction Productivity; and Total Change vs. Engineering Productivity.[178] The authors noted that the strongest statistical correlation came from two relationships: Construction Change vs. Construction Productivity and Engineering Change vs. Engineering Productivity, both of which are discussed below.

Figure 3-4 shows the correlation between engineering change and engineering productivity.[179] In the population of projects studied by CII, the vast majority required more hours to complete the original engineering scope (absent changes) than originally budgeted (PI less than 1.0). The research found that projects experiencing no engineering change still experienced a productivity index of 0.93, *i.e.*, less than planned productivity.

---

176 "Quantitative Effects of Project Change," Project Change Management Research Team, Construction Industry Institute Publication 43-2, May 1995, p. 3.

177 *Id.* at p. 7. A similar definition applies to the percent change for engineering.

178 See Ibbs, C.W. and Walter E. Allen, "Quantitative Impacts of Project Change," CII Source Document 108, May 1995, Chapter 4, p. 50.

179 *Id.* at Analysis 2, Appendix C.

Figure 3-4: Engineering Change vs. Engineering Productivity
1995 CII Study – SD 108

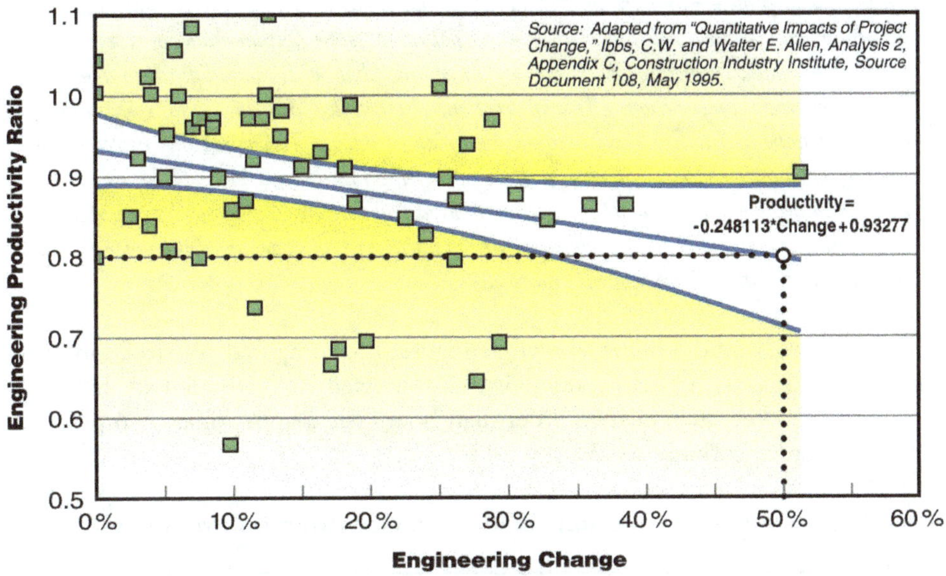

Correlation Coefficent: -0.246 @ 0.073 level Number of Observations: 54 R-Square (Regression Analysis): 0.0605
Engineering Change: Q25/Q24 Engineering Productivity: Q28

The 1995 CII Study yields an equation for estimating the Engineering Productivity Index (PI) as follows:

Engineering Productivity Index = – 0.248113 x Change + 0.93277

If the Engineering Change is 35 percent (0.35), then the equation would yield an Engineering Productivity Index as follows:

Engineering Productivity Index = – 0.248113 (0.35) + 0.93277 = 0.85

Therefore, if the contractor planned to earn 100,000 engineering man-hours, the 1995 CII Study would project the "expended" man-hours to be 100,000/0.85, or 117,647 man-hours. However, one would then need to estimate the expended man-hours if no changes had occurred, which would yield a PI = 0.93. Therefore, if no change had occurred, the study would project that the contractor would have incurred 107,527 engineering man-hours (100,000/0.93). Using the 1995 CII Study, the productivity loss is predicted to be 117,647 – 107,527 = 10,120 engineering man-hours.

The 1995 CII Study provided an explanation of these results regarding engineering, as follows:

> *This may indicate that there is less stringent cost control imposed on the detailed engineering phase of the project compared to the construction phase. Alternatively, this may reflect a willingness of project managers to expend additional funds in engineering to provide better definition and control in construction. Another possible explanation of this lower productivity index in engineering is that the concept of "change" is more difficult to assess in engineering than construction. In construction, the detailed drawings and specifications define the scope and it is generally easy to recognize a change. In engineering, however, the scope of engineering services as well as the scope of the facility tend to be uncertain at the outset and become more clear as engineering proceeds. Therefore, the baseline scope of work is difficult to establish, many changes are characterized as "design development" and consequently "changes" are difficult to define.*[180]

Another 1995 CII Study model, shown as Figure 3-5, compared construction change man-hours and construction productivity, and showed that productivity falls below the anticipated level when change order man-hours reach approximately 6 percent of total man-hour expenditures.[181]

Figure 3-5: Construction Change vs. Construction Productivity

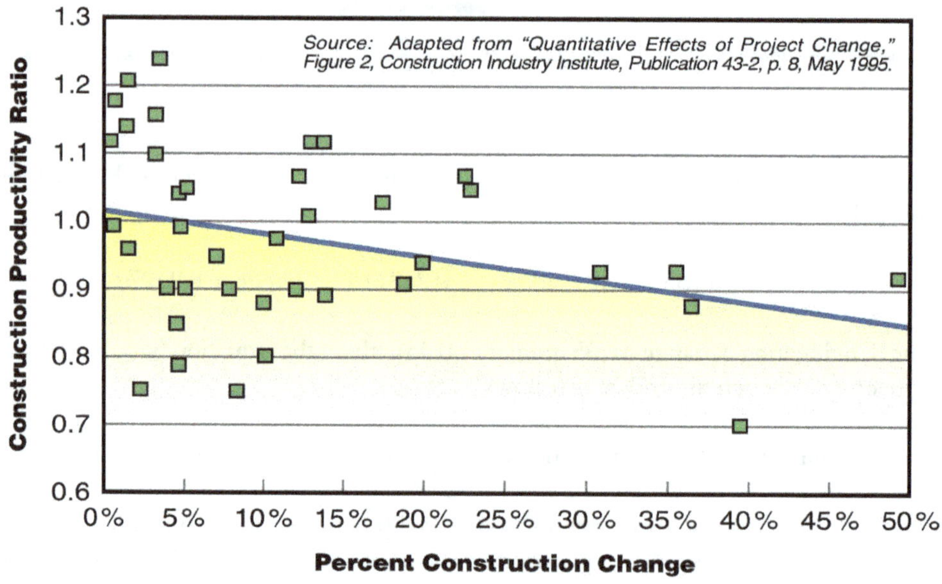

---

[180] "Quantitative Effects of Project Change," Project Change Management Research Team, Construction Industry Institute Publication 43-2, May 1995, p. 10.

[181] See "Quantitative Effects of Project Change," Project Change Management Research Team, Construction Industry Institute Publication 43-2, May 1995, p. 10. Also see Ibbs, C.W. and Walter E. Allen, "Quantitative Impacts of Project Change," CII Source Document 108, May 1995, Analysis 1, Appendix C.

The 1995 CII Study provides an example of how to use the study curves to estimate construction productivity loss, as follows:[182]

A construction project budgeted at 100,000 work-hours experienced approved scope changes amounting to 25,000 work-hours (20 percent of the total of 125,000 work-hours). Using the values in Figure 3-5, one can estimate the expected difference in final work-hours due to the addition of these 25,000 work-hours. The predicted productivity index of a project with 20 percent change is approximately 0.95. On this basis, the expected final work-hours would be 125,000/0.95,[183] or 131,600 work-hours. Had there been zero change, the predicted productivity index would be 1.02, and the expected final work-hours would be 100,000/1.02, or 98,000 work-hours. In this theoretical example, then, the additional 25,000 work-hours in scope changes would cost an additional 33,600 work-hours (131,600 – 98,000). These changes, then, cost 34 percent more to implement than if they had been included in the original scope.

Regarding construction, the 1995 CII Study yields an equation for estimating the Productivity Index (PI) as follows:

$$\text{Construction Productivity Index} = -0.343765 \times \text{Change} + 1.018244$$

If the Construction Change is 35 percent (0.35), then the equation would yield a Construction Productivity Index as follows:

$$\text{Construction Productivity Index} = -0.343765\ (0.35) + 1.018244 = 0.90$$

Therefore, if the contractor planned to earn 100,000 man-hours, the projected "expended" man-hours would be estimated to be 100,000/0.90, or 111,111 man-hours. However, one would then need to estimate the expended man-hours if no changes had occurred (based upon CII's example), which would yield a PI = 1.02. Therefore, if no change had occurred, the study would project that the contractor would have only incurred 98,039 man-hours. The productivity loss is predicted to be 111,111 – 98,039 = 13,072 man-hours.

The mean regression line crosses unity at 6 percent change, indicating that projects with less than 6 percent change, on average, experienced better productivity than planned. CII states that *"Other CII research on the impact of change orders found that the average amount of change on a different population of projects studied was also six percent. This*

182 See "Quantitative Effects of Project Change," Project Change Management Research Team, Construction Industry Institute Publication 43-2, May 1995, pp. 8–9. The values provided by CII have been rounded.

183 There is a typographical error in CII Publication 43-2, wherein it used "120,000/0.95, or 131,600 work-hours" instead of the correct "125,000/0.95, or 131,600 work-hours."

would imply that the ***productivity rates (work-hours per unit of measure) generally used in the construction industry presume that an average amount of change (six percent change orders) will be experienced****.*"[184] [emphasis added]

As described above, CII's example includes consideration for its model showing better than planned productivity in cases where there is less than 6 percent change on a project. CII concluded that for 20 percent project change, the changes "cost 34 percent more to implement than if they had been included in the original scope."[185] This does not mean that the man-hours quantified in CII's example are all claimable or recoverable, as a contractor expects some change on every project, and it should not be compensated for theoretical losses that it did not experience. For example, if there was 4 percent construction change on a project, CII's model would result in a PI of 1.004.186 While this PI is lower than the modeled value if no changes had occurred (a PI of approximately 1.2), it is still better than the contractor's planned PI of 1.0, and thus the contractor would not have experienced a loss. It follows that the "give back" in CII's example to account for no changes on a project should not be considered in the calculation of claimable man-hours.

Unlike the Leonard Study, the 1995 CII Study provided neither separate analysis of productivity loss differences between architectural/civil or electrical/mechanical projects, nor any distinction of productivity loss for other major causes. The 1995 CII Study projects less severe productivity loss than would similar data if the Leonard Study curves were used. For example, McEniry wrote the following:

- *The Leonard curves do not extend into the area where changes are less than 10% whereas the Ibbs and Allen results indicate that, in cases where changes on a project are less than 10%, productivity could in fact be even better than estimated.*

- *At 50% change, the Leonard curve for civil and architectural work would indicate a minimum loss of productivity of about 25% (changes only) while the Leonard curve for mechanical and electrical work would indicate a minimum loss of productivity of about 32%. The Ibbs and Allen curve, however, indicates a loss of productivity of only 15% regardless of the type of project.*[187]

There is a difference between the way the 1995 CII Study and Leonard measured the "percent change orders." The 1995 CII Study states:

---

184 "Quantitative Effects of Project Change," Project Change Management Research Team, Construction Industry Institute Publication 43-2, May 1995, p. 7.

185 *Id.* at p. 9.

186 PI = –0.343765 (0.04) + 1.018244 = 1.004.

187 McEniry, Gerald, "The Cumulative Effect of Change Orders on Labour Productivity – The Leonard Study 'Reloaded'," The Revay Report, Volume 26, No. 1, May 2007, p. 3.

> *The "percent change" is defined as the number of work-hours expended on authorized changes that originated during the construction phase divided by the total number of work-hours expended for construction.*[188]

The 1995 CII construction study does not indicate, as did Leonard, that the denominator in this case excludes the change order hours, although it would be logical to have done so. It also seems uncertain whether any evaluation was done by the CII construction study to adjust for contractor inefficiencies.

Criticism of the 1995 CII construction study includes the following:

- The scatter in the data and consequent low correlation between the productivity index and the "percent change orders," as can be seen in Figure 3-5;[189]
- Data on projects affected by greater than 25 percent change was sparse, and the alignment of an extrapolated trend line could be dramatically affected by a few distant points;[190] and
- This study has not yet been discussed in U.S. case law or board decisions.[191]

The authors of the 1995 CII Study also indicated that *"Although the productivity charts in this publication are too general to be used in all circumstances to price every individual change precisely, they should give a project manager a sense of what to look for on projects that are being disrupted by many changes."*[192]

---

188 "Quantitative Effects of Project Change," Project Change Management Research Team, Construction Industry Institute Publication 43-2, May 1995, p. 7.

189 See McEniry, Gerald, "The Cumulative Effect of Change Orders on Labour Productivity – The Leonard Study 'Reloaded'," The Revay Report, Volume 26, No. 1, May 2007, p. 3.

190 See Ibbs, William and Gerald McEniry, CFCC PSP, "Evaluating the Cumulative Impact of Changes on Labor Productivity – an Evolving Discussion," Cost Engineering, Association for the Advancement of Cost Engineering International, Vol. 50/No. 12, December 2008, pp. 23–29.

191 See McEniry, Gerald, "The Cumulative Effect of Change Orders on Labour Productivity – The Leonard Study 'Reloaded'," The Revay Report, Volume 26, No. 1, May 2007, p. 4.

192 "Quantitative Effects of Project Change," Project Change Management Research Team, Construction Industry Institute Publication 43-2, May 1995, p. 12.

The authors of the 1995 CII Study concluded: *"This research confirms that project change has a sizable impact on project productivity, schedule and costs."*[193] The analysis includes several models, similar to those presented by Leonard, which show a direct relationship between change order man-hours and loss of productivity.[194]

The 1995 CII Study also discussed the impact of late changes on schedule and budget recovery:

> *Although the industry acknowledges that the later a change occurs on a project the less efficiently it is implemented, many projects execute significant amounts of changes late in project life cycles. This is especially counter-productive since it can be demonstrated that projects have a significantly reduced ability to recover schedule losses or budget pressures as they approach completion.*[195]

It is therefore logical that the timing of changes may also have a significant bearing on a cumulative impact of changes claim. Changes approved very early in the contract before field management, labor supervision, and direct and subcontract labor have been mobilized to the site have less impact or no impact at all, compared to the same changes occurring in the middle or late in the project. The later into the project engineering and construction changes are made, the more impact those changes have on project execution time and cost. Not only is the impact that each individual change exerts on project cost and schedule greater when engineering and construction are more complete, but also the impact is greater when there is a large number of changes. A quantitative analysis of productivity loss based on the timing of changes is addressed in Section 3.4.1.3.5.

3.4.1.3.4 2001 CII Study

Another CII study submitted in 2001 attempted to *"quantify the cumulative impact that change orders have on the overall project productivity, including the effect on the base contract, for mechanical and electrical construction."*[196] This study examined data from 116 projects[197] and provided an equation *"to predict the impact of change orders on labor productivity."*[198]

---

193 Ibbs, C.W. and Walter E. Allen, "Quantitative Impacts of Project Change," CII Source Document 108, May 1995, Chapter 6, p. 76.

194 *Id.* at Appendix C.

195 "Quantitative Effects of Project Change," Project Change Management Research Team, Construction Industry Institute Publication 43-2, May 1995, p. 25.

196 Hanna, Awad S., "Quantifying the Cumulative Impact of Change Orders for Electrical and Mechanical Contractors," CII Research Report 158-11, February 2001.

197 *Id.* at p. 133.

198 *Id.* at p. 108.

Underlying the 2001 CII Study were two 1999 studies performed by research teams at the University of Wisconsin.[199]

For the analysis of mechanical construction productivity, data from 61 projects were compiled with an average project size of 22,268 man-hours. The average change orders on these projects, measured in labor hours as a percent of estimated base hours, was 22.8 percent.[200] For the analysis of electrical construction productivity, data from 61 projects were compiled with an average project size of 18,000 man-hours.[201]

The 2001 CII Study provided an equation *"to predict the impact of change orders on labor productivity."*[202] The equation includes six factors which affect the loss of productivity estimation, specifically: (i) percentage of change order man-hours to budgeted man-hours; (ii) percentage of time spent on the project by the project manager; (iii) percentage of change orders initiated by the owner; (iv) whether or not productivity was tracked during the project; (v) whether or not the project was overmanned; and (vi) the average processing time of the change orders.[203]

$$\begin{gathered}\%\text{Delta} = 0.36866 + 0.11957\ \text{Percent Change} - \\ 0.08065\ \text{PM\%TimeOnProject} - 0.16723\ \%\text{OwnerInitiatedCO} - \\ 0.09147\ \text{Productivity} - 0.05213\ \text{Overmanning} + \\ 0.022345\ \text{Processing Time}\end{gathered}$$

199 See Hanna, Awad S., Jeffrey S. Russell, Timothy W. Gotzion, and Erik V. Nordheim, "Impact of Change Orders on Labor Efficiency for Mechanical Construction," Journal of Construction Engineering and Management, ASCE 125: 3 (1999b), pp. 176–184; and Hanna, Awad S. Jeffrey S. Russell, Erik V. Nordheim, and Matthew J. Bruggink, "Impact of Change Orders on Labor Efficiency for Electrical Construction," Journal of Construction Engineering and Management, ASCE 125: 4 (1999b), pp. 224–232. Because only 116 projects were used in the 2001 CII Study, six projects must have been discarded from the separate mechanical and electrical productivity analyses.

200 See Hanna, Awad S., Jeffrey S. Russell, Timothy W. Gotzion, and Erik V. Nordheim, "Impact of Change Orders on Labor Efficiency for Mechanical Construction," Journal of Construction Engineering and Management, ASCE 125: 3 (1999b), p. 178.

201 See Hanna, Awad S. Jeffrey S. Russell, Erik V. Nordheim, and Matthew J. Bruggink, "Impact of Change Orders on Labor Efficiency for Electrical Construction," Journal of Construction Engineering and Management, ASCE 125: 4 (1999b), p. 228.

202 Hanna, Awad S., "Quantifying the Cumulative Impact of Change Orders for Electrical and Mechanical Contractors," Construction Industry Institute Research Report 158-11, February 2001, p. 108.

203 *Id.*

The authors define "Delta" as the difference between the actual labor hours expended to complete the project, and the estimated base hours plus change order hours.[204] Positive values of "Delta" indicate that the actual productivity is less than the planned or estimated productivity. Positive "Delta" can be attributed to a variety of factors, such as the contractor's low estimate, a contractor's inefficiency, and the impact of productivity-related factors such as change orders, weather conditions, work interruptions, and rework, among others. Projects that were impacted by other factors or impacted because of the contractors' low estimates were eliminated from the database.[205]

Ibbs pointed out the limitations of the Hanna study, as follows:

> *There are several limitations with this study, one being the use of the "impact classification" variable. This variable was qualitatively assessed by the contractors responding to the survey. Those contractors, who have a vested interest in the results of the study, ranked the degree to which their projects were affected by factors such as change orders, work sequence disruption, trade stacking, material problems, weather conditions, overtime, and shift work. Not all of these variables are the responsibility of the owner, though the model presumes such. Weather risks—as an example—may be contractually allocated to the contractor. Moreover, the ranking was done on a subjective 0–5 scale, "less" to "more." The writers do not explain how they controlled for this subjectivity.*
>
> *Another unclear aspect is that the number of change orders is a significant variable in the mechanical. Thus the same job could have two different predicted productivity losses, depending only on whether an owner approves a particular change order request, not on the merits per se of the change.*[206]

McEniry criticized the 2001 CII Study and the underlying Hanna studies, as follows:

- Estimates of the overall project hours and the change order hours were assumed to be reliable.
- The actual change order hours used in the studies were not tracked, but rather estimated.
- The proposed relationship of the number of variables to be evaluated is complex, with different variables for each trade.

---

204 See Hanna, Awad S., Jeffrey S. Russell, Timothy W. Gotzion, and Erik V. Nordheim, "Impact of Change Orders on Labor Efficiency for Mechanical Construction," Journal of Construction Engineering and Management, ASCE 125: 3 (1999b), pp. 176–184.

205 *Id.* at p. 177.

206 Ibbs, William, "Impact of Change's Timing on Labor Productivity," Journal of Construction Engineering and Management, ASCE 131:11, 2005a, p. 1221.

- Some of the variables give counterintuitive results. For instance, it would seem that lost productivity would be less if the project manager spent more time on the project and over-manning occurred.

- The ranking of the impact classification, the only variable common to both electrical and mechanical work, is considered subjective.[207]

The authors are not aware of any instances in which these studies have been endorsed in U.S. case law or board decisions published to date.

3.4.1.3.5 2005 Ibbs Study

In December 2008, AACE International published a paper entitled "Evaluating the Cumulative Impact of Changes on Labor Productivity – an Evolving Discussion."[208] Dr. Ibbs performed studies in 1995,[209] 1997,[210] and 2005,[211] based on data compiled from research studies of actual construction projects. These studies developed a statistical relationship between the amount of change and end-of-project labor productivity.

Including the 1995 CII Study, Dr. Ibbs collected data from owners and contractors from 170 projects over 12 years,[212] ranging in size from $2 million to $14 billion, with a median value of $44 million. The projects included both public and private projects with different delivery systems. The data includes change and productivity information for both the design and construction phases of the projects.

The 1995 CII Study reported that no meaningful change-productivity differences existed for architectural/civil vs. electrical/mechanical projects, but the study did evaluate the impact of change orders on both engineering and construction

---

207 See McEniry, Gerald, "The Cumulative Effect of Change Orders on Labour Productivity – The Leonard Study 'Reloaded'," The Revay Report, Volume 26, No. 1, May 2007, p. 4.

208 See Ibbs, William and Gerald McEniry, CFCC PSP, "Evaluating the Cumulative Impact of Changes on Labor Productivity – an Evolving Discussion," Cost Engineering, AACE International, Vol. 50/No. 12, December 2008. Also see Ibbs, William, "Impact of Change's Timing on Labor Productivity," Journal of Construction Engineering and Management, ASCE 131:11, 2005a, pp. 1219–1223.

209 See Ibbs, William C., "Quantitative Impacts of Construction Change: Size Issues," ASCE Journal of Construction Engineering and Management, 123, 3, 1997, pp. 308–311.

210 See Ibbs, C.W. and Walter E. Allen, "Quantitative Impacts of Project Change," CII Source Document 108, May 1995.

211 See Ibbs, William, "Impact of Change's Timing on Labor Productivity," Journal of Construction Engineering and Management, ASCE 131:11, 2005a, pp. 1219–1223.

212 The 1995 CII Study contained 104 projects.

productivity. Ibbs' 2005 study collected many more data points in the range of 25 to 50 percent change orders, but he concentrated his 2005 study on construction productivity. The percent change is thus based on construction labor-hours, and the Productivity Index is equal to actual productivity/planned productivity, where a value above 1.0 indicates performance better than expected, and the actual productivity is based on end-of-project productivity values.[213]

Figure 3-6 presents a chart from Ibbs' and McEniry's 2008 article,[214] which compares Ibbs' data from the 1995 CII Study (Ibbs and Allen)[215] and his 2005 data.[216]

Figure 3-6: Comparison of Ibbs' 1995 and 2005 Productivity Loss Data

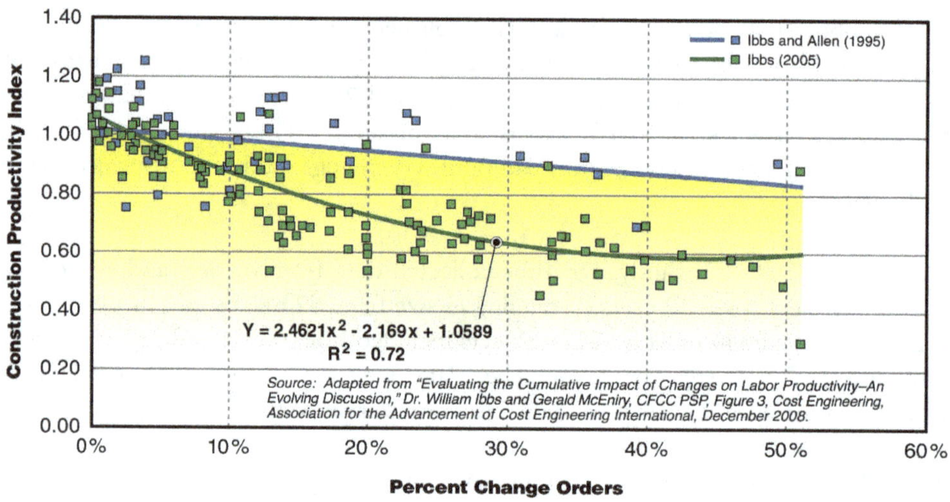

To use this chart, Ibbs presented a formula for relating change orders and Construction Productivity Index, using a definition from research conducted by Charles A. Leonard in 1988.[217] The formula is stated below:

---

213 See Ibbs, William, "Impact of Change's Timing on Labor Productivity," Journal of Construction Engineering and Management, ASCE 131:11, 2005a, p. 1221.

214 See Ibbs, William and Gerald McEniry, CFCC PSP, "Evaluating the Cumulative Impact of Changes on Labor Productivity – an Evolving Discussion," Cost Engineering, AACE International, Vol. 50/No. 12, December 2008, Figure 3.

215 See Ibbs, C.W. and Walter E. Allen, "Quantitative Impacts of Project Change," CII Source Document 108, May 1995.

216 See Ibbs, William, "Impact of Change's Timing on Labor Productivity," Journal of Construction Engineering and Management, ASCE 131:11, 2005a, p. 1221.

217 See Leonard, Charles A., "The Effects of Change Orders on Productivity," M.S. Thesis at Concordia University, 1988. Also see Chapter 3, Section 3.4.1.3.1 herein.

Percent change orders = change order labor hours[218]/(total actual labor-hours – change order hours – contractor mistakes)

Using this definition, Ibbs also presented a formula which correlated "% Change Orders" to "Construction Productivity Index," or PI, where the Loss of Productivity (LOP) is defined as follows:

LOP = (1 – PI) x 100%, where PI = earned hours/actual hours

Based on Ibbs' chart and the two formulae presented above, one can estimate the Productivity Index for a given Percent Change that occurred in a project. For example, if the Percent Change Orders was 35 percent (x = 0.35), using Ibbs' formula, one would derive the following estimate of loss of productivity:

$PI = 2.4621x^2 – 2.169x + 1.0589$

$PI = 2.4621 \times (0.35)^2 – 2.169 (0.35) + 1.0589$

PI = 0.3016 – 0.7592 + 1.0589

PI = 0.6013

LOP = (1 – 0.6013) x 100% = 39.87 or approximately 40%

For this same value of 35 Percent Change Orders, Ibbs' 1995 data from Figure 3-5 above would have yielded only a 10 percent Loss of Productivity. Thus, Ibbs' 2005 data indicates that changes had a more significant impact on productivity than indicated in the 1995 CII Study

If no change occurred, Ibbs' 2005 data would indicate that productivity would be 6 percent better than planned, *i.e.*, the contractor would have expended fewer man-hours than it expected. At 2.8 percent change, the Productivity Index was 1.0, meaning that the actual man-hours were equal to the planned man-hours. Productivity was negatively affected as changes increased above 2.8 percent.

In comparing the results of the Ibbs 2005 study and the 1988 Leonard Study, McEniry highlights another problem measuring "percent change orders" and lost productivity, as follows:

---

218 Leonard concludes that actual hours spent by the contractor on change order work are usually not available. Therefore, earned change order hours, obtained from estimates, are utilized for change order labor hours. See Leonard, Charles A., "The Effects of Change Orders on Productivity," M.S. Thesis at Concordia University, 1988, p. 48.

> *Typically, change order hours are hours paid for authorized extra work, while the amount of extras so authorized is dependent upon the project team's ability to have alleged extra work recognized and paid as such. Obviously, not all contractor claims for extra work will be recognized. Some of the additional hours worked may well be the result of disputed extra work and some will be the impact of the authorized changes. Practically, it may be difficult to distinguish between the two and the situation could give rise to different interpretations. For instance, on a given project there was a theoretical 30% increase in scope. One team may get paid for 40% as the increased scope while another team may get paid for only 20%. Based on these numbers the team getting paid only 20% would appear to have suffered a more significant loss of productivity considering the actual amount of unpaid work hours.*[219]

McEniry's point is a valid one and applicable not only to Ibbs' 2005 study, but to other studies which utilize authorized or owner-approved change order man-hours in their quantification of productivity loss. Section 3.4.1.3.7 further discusses the problems of using these studies in practice, especially in respect to projects that have a substantial value of disputed changes.

In addition, Ibbs discusses that both additive changes and deductive changes should be used to determine the amount of change, as follows:

> *Change is measured in absolute terms, meaning that a project that [has] $1 of deductive changes and $1 of additive change would be treated as having $2 of total change. The rationale is that deductive change can be disruptive to productivity, just as additive change can be. To use the net difference between additive and deductive changes would understate that disruption. For example, netting the difference between $1 of additive and $1 of deductive change would indicate that there was no change even though there were probably productivity perturbations.*[220]

Ibbs also presented his 2005 impact results as a function of the timing of change. He determined a regression equation for late, normal, and early change, which Ibbs defined as follows:

> Late Change = 50 percent of the changes were recognized when the project was 70 percent complete[221]

---

[219] McEniry, Gerald, "The Cumulative Effect of Change Orders on Labour Productivity – The Leonard Study 'Reloaded'," The Revay Report, Volume 26, No. 1, May 2007, p. 6.

[220] Ibbs, William, "Impact of Change's Timing on Labor Productivity," Journal of Construction Engineering and Management, ASCE 131:11, 2005a, p. 1221.

[221] Based on author's communications with Dr. Ibbs, because this analysis was based only on construction productivity, the project start is defined to be construction start. Therefore, 20 percent complete means 20 percent complete with construction work hours as measured by Earned Value.

Normal Change = 50 percent of the changes were recognized when the project was 40 percent complete

Early Change = 50 percent of the changes were recognized when the project was 20 percent complete[222]

Regarding recognition of changes, Ibbs states:

> *Formal recognition in this case is when one party notifies the other by either verbal or written notice, whichever is earlier. There is some imprecision in this definition but this researcher felt it was the best choice available.*[223]

Also, because productivity is based on cumulative, end-of-project productivity values, late changes have fewer hours to impact. Therefore, Ibbs' analysis shows that late changes are extremely disruptive to the man-hours remaining in the later stages of the project, because that disruption is powerful enough to drag the cumulative productivity value down under the other two curves. Conversely, it is easier to recover from early changes, because there is time to make up for lost productivity.[224]

Figure 3-7 recreates a graphic (Figure 5) from Ibbs' and McEniry's 2008 article,[225] which includes his regression curves and equations from his 2005 data for late, normal, and early changes.

---

222 See Ibbs, William and McEniry, Gerald, "Evaluating the Cumulative Impact of Changes on Labor Productivity – an Evolving Discussion," Cost Engineering, AACE International, Vol. 50/No. 12, December 2008, p. 26.

223 Ibbs, William, "Impact of Change's Timing on Labor Productivity," Journal of Construction Engineering and Management, ASCE 131:11, 2005a, p. 1221.

224 *Id.* at p. 1222.

225 See Ibbs, William and McEniry, Gerald, "Evaluating the Cumulative Impact of Changes on Labor Productivity – an Evolving Discussion," Cost Engineering, AACE International, Vol. 50/No. 12, December 2008.

Figure 3-7: Ibbs' Productivity Loss Data as a Function of the Timing of Changes.

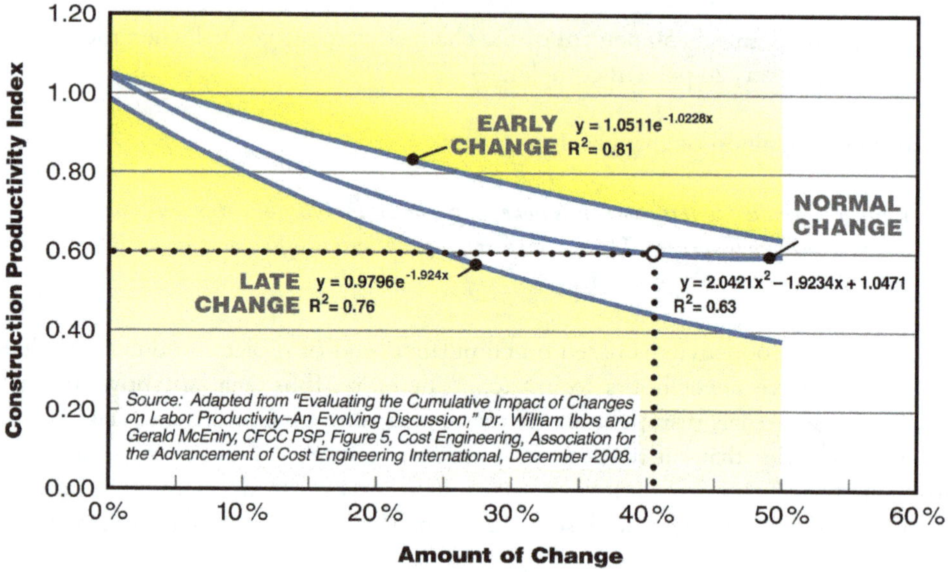

The three regression equations derived from Ibbs' 2005 data are listed below:

Early Change: $PI = 1.0511e^{-1.0228x}$

Normal Change: $PI = 2.0421x^2 - 1.9234x + 1.0471$

Late Change: $PI = 0.9796e^{-1.924x}$

Therefore, using the three equations above, if 50 percent of the changes were recognized when the project was 20 percent complete (Early Change), Ibbs' data would predict about a 23 percent loss of productivity if the Percent Change was 30 percent (x = 0.30), as follows:

Early Change: $PI = 1.0511e^{-1.0228x}$

$PI = 1.0511e^{-1.0228(.30)}$

$PI = 0.7734$

Loss of Productivity = (1 – 0.7734) x 100% = 22.66%

However, if 50 percent of the changes were recognized when the project was 40 percent complete (Normal Change), Ibbs' data would predict about a 34 percent loss of productivity if the Percent Change was 30 percent (x = 0.30), as follows:

Normal Change: $PI = 2.0421x^2 - 1.9234x + 1.0471$

$PI = 2.0421\ (.30)^2 - 1.9234(0.30) + 1.0471$

$PI = 0.1838 + 0.5770 + 1.0471$

$PI = 0.6539$

Loss of Productivity = (1 – 0.6539) x 100% = 34.41%

Finally, if 50 percent of the changes were recognized when the project was 70 percent complete (Late Change), Ibbs' data would predict a 45 percent loss of productivity if the Percent Change was 30 percent (x = 0.30), as follows:

Late Change: $PI = 0.9796e^{-1.924x}$

$PI = 0.9796e^{-1.924(0.30)}$

$PI = 0.55$

Loss of Productivity = (1 – 0.55) x 100% = 45%

Thus, the 2005 Ibbs study would predict that projects with late changes experience much more disruption than projects where change is recognized earlier.

The study assumes that: (i) the contractor's estimated man-hours for the original work and the change order work are accurate, and (ii) the actual man-hours include only loss of productivity due to change orders, and not due to contractor inefficiency. According to Dr. Ibbs, "*…these are major assumptions and could easily and substantially skew the results if they were not accurate representations of the project conditions.*"[226]

The authors are not aware of cases where this research has been cited in published U.S. case law, or in the case law of other jurisdictions.

3.4.1.3.6 Court's General View of Formulaic Approaches

With the exception of the Leonard Study, published court and board decisions do not provide a view on the applicability of cumulative impact studies. However, in *Aetna Casualty & Surety Co.*, the board stated that formulaic approaches are generally viewed with skepticism:

---

226 Ibbs, William, "Impact of Change's Timing on Labor Productivity," Journal of Construction Engineering and Management, ASCE 131:11, 2005a, p. 1221.

> *Formulaic approaches to proving productivity losses are viewed with skepticism by the courts. Studies, formulas and other information relied upon by experts must be shown relevant and applicable to the Project at issue. As explained in McGee Landscaping, Inc...in order for a formula, study, textual material or handbook to be used to determine lost productivity, there must be specific evidence of the applicability of the manual's underlying assumptions to the situation at hand. Where this necessary evidence is not presented, a claim will be rejected for insufficient information to calculate the loss, even if the trier of fact believes that there was some loss of productivity.*[227]

Given the court's unfavorable view of the Leonard Study[228] and the above-cited questions of the applicability of formulaic approaches, the use of cumulative impact studies is not the most favorable method to estimate productivity loss. However, published decisions do not tell the whole story. Ibbs has written that he has successfully used both Leonard and his own studies in mediation, arbitration, and court settings.[229]

3.4.1.3.7 Further Uncertainties in using Formulaic Approaches

Several of the methods used to estimate disruption caused by changes utilize authorized or owner-approved change order man-hours, including the 1988 Leonard Study, the 1995 CII Study, the 2001 CII Study, and the 2005 Ibbs Study, as described above. In each study, the researchers collected data from numerous projects, including the authorized change order man-hours and the overall labor productivity on those projects. However, the studies present no information with regard to the magnitude of disputed change order man-hours or the composition of man-hours included in the agreed change orders (*i.e.*, direct extra work versus direct disruption man-hours). Moreover, the proper handling of direct disruption is not clear: should direct disruption be treated as a project change and handled as an input into the formulae, or is direct disruption included in the lost man-hours from the formulae's results?

This lack of definition is problematic when attempting to assess loss of productivity on a given project, especially if the project is highly impacted and the relationship between the contractor and the owner is contentious. On such a project, it would not be unusual for the owner to have authorized change orders totaling 10,000 man-hours, but the parties to also have disputed change orders comprising

---

227 *Aetna Casualty & Surety Co. v. George Hyman Construction Co., U.S. Dist.,* LEXIS 22627 (E.D. Pa. 1998), pp. 273–274 (citing *McGee Landscaping, Inc.,* AG-BCA No. 91-172-1, 93-3 BCA (CCH) ¶ 25,946 (1993) and *Groves-Black,* ENGBCA No. 4557, 85-3 BCA (CCH) ¶ 18,398 (1985).

228 See Section 3.4.1.3.1 herein.

229 See Ibbs, "The Cumulative Impact of Change on Construction Labor Productivity," 2008, p. 4.

20,000 man-hours, and the contractor to also have a direct disruption claim totaling 40,000 man-hours. Using the studies strictly as they were intended, the contractor should only be able to estimate lost man-hours using the authorized change orders, or 10,000 man-hours. However, this would yield far less cumulative impact than if the contractor used the man-hours that, in its opinion, the owner should have approved. The question of how many man-hours to use – 10,000 man-hours approved, 30,000 man-hours including disputed changes, or 70,000 man-hours including the direct disruption claim – is dependent upon how the project aligns with the projects in the particular study, which cannot be known. There is no clear answer to the issues raised above, and the lack of definition in the studies introduces another level of uncertainty when using these studies to estimate disruption.

As a second point, McEniry (in reference to the 2005 Ibbs Study) recommended applying the calculated loss of productivity only to those man-hours that might have been impacted:

> *Moreover, it should be realized that estimated factors of lost productivity derived from the above graphics should not be applied blindly across the whole project. An effort must be made to determine when the changes took (or will take) place and apply the estimated loss of productivity factor to the hours worked (or to be worked) in a particular period. For instance, if most changes were initiated on a specific project between 50 and 100% completion, the estimated loss of productivity should not be applied to man-hours worked before this period, unless it can be demonstrated that the changes had some detrimental effect.*[230]

McEniry's point has some merit, as one can imagine a scenario wherein a majority of the changes occurred very late in a project, the calculated loss of productivity was applied to all actual man-hours, and the resulting lost man-hours exceeded the actual man-hours that the contractor could have lost late in the project.

However, any attempt to subdivide a project by timeframe or scope of work may be a deviation from the underlying data in the industry studies. In the studies, the relationships between the magnitude of changes and productivities were developed based upon comparisons between total project change and total project productivity. Therefore, using the studies to quantify productivity losses on one component of a given project, such as the mechanical works alone or the last half of a given project, may be inappropriate and will likely be met with criticism.

230 McEniry, Gerald, "The Cumulative Effect of Change Orders on Labour Productivity – The Leonard Study 'Reloaded'," The Revay Report, Volume 26, No. 1, May 2007, p. 7.

3.4.1.4 The Use of Factors

Productivity loss is often estimated using disruption factors applied to planned or actual man-hours. Below, the MCAA Factors and the NAVSEA Guidelines are discussed.

3.4.1.4.1 MCAA Factors

The Mechanical Contractors Association of America (MCAA) issued Bulletin No. 58 in 1976, entitled "Factors Affecting Productivity." This document lists sixteen factors which potentially impact productivity and provides a range of productivity losses for each factor, dependent on the severity of the conditions. The factors include: *Stacking of Trades; Morale and Attitude; Reassignment of Manpower; Crew Size Efficiency; Concurrent Operations; Dilution of Supervision; Learning Curve; Errors and Omissions; Beneficial Occupancy; Joint Occupancy; Site Access; Logistics; Fatigue; Ripple; Overtime;* and *Season and Weather Change.*[231]

In 2005, the MCAA issued a document, "Change Orders Productivity Overtime: A Primer for the Construction Industry." This document was again updated in 2011 and 2012. In the 2012 document, the MCAA includes a table with the same sixteen "Factors Affecting Labor Productivity."[232] The factor titles, descriptions, and estimated impacts have remained unchanged from the MCAA's earlier publications.

The MCAA advises that *"the specific values shown in the factor tables must be applied with careful consideration and a review of the facts surrounding the events, which caused the loss of productivity. The applications of the various MCAA factor percentages will vary as project conditions dictate."*[233]

3.4.1.4.1.1 General Considerations

In its 2012 publication, the MCAA makes certain important commentary regarding productivity loss claims, as follows:

- *"Once the mechanical contractor's estimate has been eliminated as the source of the loss, the mechanical contractor should determine what other events cause the loss of productivity."*[234] Thus, the MCAA recognizes that the contractor's bid estimate must be reviewed before a productivity loss claim can be properly prepared.

---

231 "Factors Affecting Productivity," Mechanical Contractors Association of America, Bulletin No. 58, 1976, was superseded by MCAA in its publication "Change Orders Productivity Overtime – A Primer for the Construction Industry," which was issued in 2005, and updated in 2011 and 2012, pp. 77–78.

232 *Id.* pp. 77–78.

233 *Id.* p. 81.

234 *Id.* at pp. 73–74.

- "Reservation of rights" or "accord and satisfaction" language must be addressed before a productivity loss claim can be upheld.[235]

- *"Factors Affecting Labor Productivity"* include three levels of potential productivity impacts: "Minor," "Average," and "Severe." Each level of impact intensity carries its own loss of productivity percentage. The three impact levels indicate the estimated effects of the changed condition on the labor hours being analyzed, *i.e.*, specific hours within the total hours expended, or on the total hours expended on the project, depending on the approach being used. Also, the three levels of intensity allow the user to more specifically assign an estimated impact for each of the MCAA factor categories being used, and like the categories themselves, should be *"**applied with care and, if at all possible, with input by those who witnessed the conditions under evaluation**."*[236] [emphasis added]

- The MCAA cautions about the duplicative use of factors, as follows:

> *... care must be taken to eliminate overlapping factors, to the fullest extent possible. The unrestrained and ill considered choice of multiple factors can lead to unreliable results.*
>
> *For instance, the factor describing "Morale and Attitude" is a valid, but somewhat amorphous, category of inefficiency. The effects of a decline in workplace morale and attitude can be embodied in several other MCAA factors, such as stacking of trades, overtime fatigue and reassignment of manpower. It would be impossible to determine what portion the impact percentage caused by stacking of trades, overtime fatigue and reassignment of manpower is attributable to the attendant decrease in worker morale and attitude. Thus, by way of the above example, when using other factors that may already include in the loss of productivity factor a consideration for decreased worker morale and attitude, it may be advisable to avoid applying a potentially duplicative factor such as "Morale and Attitude."*
>
> *Another example to consider when striving to avoid factor duplication is the "Ripple Effect." This term of art has been used in several board decisions and is a well recognized phenomenon in the construction industry. This MCAA factor describes the downstream effect on the mechanical contractor of impacts caused to predecessor trades. For example, the mechanical contractor's schedule may be compressed because the building structure was erected late. In order to mitigate the structural delay, the general trade contractor may*

235 *Id.* at pp. 13, 38, 40, 56 (footnote 8), 58, 84, 85, 98, and 100–101.

236 *Id.* at p. 87.

> *accelerate the follow on trades by stacking the crews of the various subcontractors, or forcing the subcontractors to work on an overtime basis. In such cases, the loss of productivity may be better classified by the events that result from the ripple effect, such as "Stacking of Trades" or "Overtime Fatigue."*
>
> . . .
>
> *Indiscriminate assignment of the MCAA factors can result in estimates that may be overstated and unreliable."*[237]

- The MCAA cautions that the intensity percentage must be chosen with care:

> *. . . if the change in scope is of a limited nature, on a project with a reasonably small crew size with little or no schedule impact (as opposed to productivity impact), then a "minor" category can be chosen. However, if the change is significant in its scope and requires major rescheduling and/or resequencing, crew size increases, overtime, shifting of work areas, piecemealing of the work and general disruption of the rhythm of the crews, then "average" or "severe" impacts could be the result.*[238]

- The MCAA recommends that an estimated/planned versus actual craft curve is appropriate for depicting the effects of unplanned crew size growth.[239]

Care must be taken in using the MCAA factors in estimating loss of productivity man-hours. Specifically, the MCAA states:

> *The percent values for each category chosen are additive in the change order pricing. Once all of the factors have been carefully evaluated for each changed condition caused solely by the proposed change in scope,* ***the percentages are added together. The total percent is then multiplied against the estimated craft labor hours for the change.***[240] [emphasis added]

3.4.1.4.1.2 Method

The MCAA method provides a means to either estimate productivity losses based on utilizing either planned man-hours (*i.e.*, forward pricing) or actual man-hours (*i.e.*, retrospectively). Either approach must be performed by proper application of the

---

237 *Id.* at pp. 87–88.

238 *Id.* at p. 88.

239 *Id.* at pp. 88–89.

240 *Id.* at p. 89.

method, and with proper consideration of the project-specific facts. The MCAA states:

> *The "time specific" method is used for both this modified forward pricing method and the retroactive pricing method (with slightly different rules)...*[241]

A sample application of the forward pricing method is provided by the MCAA,[242] as shown in Table 3-5:

Table 3-5: Example of Forward Pricing Productivity Loss Estimations

Change Order Impact Period	**Week 1**	**Week 2**	**Week 3**	**Week 4**	**Week 5**	**Week 6**
Original Planned Hours	400	400	600	600	**720**	800
Estimated C.O. Hours	80	160	320	400	400	160
Revised Planned Hours	**480**	**560**	**920**	**1,000**	**1,120**	**960**
Learning Curve	5%	5%	5%	0%	0%	0%
Dilution of Supervision	0%	10%	10%	10%	10%	10%
Crew Size Inefficiency	0%	0%	10%	10%	10%	0%
Total MCAA Factor	**5%**	**15%**	**25%**	**20%**	**20%**	**10%**
Estimated Loss of Productivity	**24**	**84**	**230**	**200**	**224**	**96**
					Total	**858**

With regard to the above example, the MCAA further states:

> *In this example, a specific time frame has been evaluated for estimated impacts. This more specific method permits the contractor to make MCAA factor applications, which can vary as estimated conditions vary. This is actually more realistic and compares well with what actually happens in the field when changes are issued, or when acceleration or other impacts occur. In reality, as time and conditions in the field change, the MCAA factors can change as well and the estimate should reflect this fact.*[243]

The MCAA also identifies certain considerations that "*must be made*" in retrospectively preparing a loss of productivity claim, including:

---

241 *Id.* at p. 90.

242 *Id.* at pp. 90–91.

243 *Id.* at p. 91.

Before a contractor makes a claim for a loss of labor productivity at the conclusion of a project, several obvious considerations must be made, including:

1. *Was the estimate/plan of craft hours accurate and reasonable?*
2. *Were the conditions, which caused the loss of productivity, reasonably foreseeable when the project was bid/negotiated?*
3. *Did the contractor cause this loss of productivity?*
4. *Were the principal causes for the loss of productivity the responsibility of identifiable parties?*
5. *Will the potential cost of recovery exceed the loss?* [244]

These considerations would require anyone who is evaluating the validity of the contractor's claim to have the following information:

- The contractor's bid estimate in sufficient specificity to evaluate the reasonableness and accuracy of the estimated craft hours by discipline; and

- An evaluation of the contractor's responsibility for the loss of productivity,[245] including any sharing of responsibility for any of the impact factors for which the owner may be responsible. For example, trade stacking may be caused by delay for which both the owner and contractor are responsible.

Regarding the actual productivity loss estimation method, the MCAA states:

> *Often, the method used by contractors is to multiply the cumulative percentage of losses of productivity as derived from the MCAA factors against the total, actual hours expended, sometimes with, and sometimes without, change order hours included in the total. This methodology of multiplying the MCAA factor percent against the actual hours is incorrect. The actual hours against which the MCAA factors are frequently multiplied in a contractor's retroactively priced claim for loss of productivity already include the contractor's loss of productivity; therefore multiplying the MCAA factors against the actual hours overstates the loss of productivity.* ***Only by removing the theoretically efficient hours from the contractor's actual hours can the MCAA factors be properly applied in a retroactively priced request for equitable adjustment****.* [emphasis added]
>
> *The actual hours must be further adjusted to deduct:*

---

244 *Id.* at pp. 92–93.

245 See an identification of certain contractor-caused problems in Chapter 3, Section 3.2 herein.

1. *Time and materials hours;*
2. *Hours spent to repair the contractor's defective work;*
3. *Change orders on which a loss of productivity has already been calculated; (If the contractor has included "forward priced" loss of productivity in individual, executed change orders, and then seeks to recover global losses at the end of the project, these incremental, per change order loss estimates must be factored out of the computations.)*
4. *Hours associated with executed change orders, where it has been determined that the contractor is barred from recovering the impact caused by the executed change orders;*
5. *Hours expended by crews that were not affected by a loss of productivity;*
6. *Other types of productivity losses for which the contractor is responsible (i.e., bid errors).*[246]

Care must also be taken to properly apply the factors to the appropriate time frame of the work that was affected. The MCAA states:

> *The MCAA factor percentages sometimes change as actual project conditions change. Therefore, it can be useful to assign the MCAA factors to the specific impacted time frames within the overall project duration. In some cases, multiplying an MCAA factor against the total hours expended for the total duration of the project will result in a distortion (on the high side) of the forecasted loss of productivity.*[247]

A sample application of a retrospective assessment is provided by the MCAA,[248] as shown in Table 3-6.

246 "Change Orders Productivity Overtime – A Primer for the Construction Industry," MCAA, 2012, pp. 93–94.

247 *Id.* at p. 94.

248 *Id.* at p. 95.

Table 3-6: Example of Retrospective Productivity Loss Estimations

Contract Period	**Week 40**	**Week 41**	**Week 42**	**Week 43**	**Week 44**	**Week 45**
Actual Payroll Hours	1,600	1,600	1,800	2,400	2,400	3,200
Deducted Hours	0	-80	-120	0	-120	-120
Revised Actual Hours	**1,600**	**1,520**	**1,680**	**2,400**	**2,280**	**3,080**
Reassignment of Manpower	5%	5%	5%	10%	10%	10%
Dilution of Supervision	0%	10%	10%	10%	10%	10%
Crew Size Inefficiency	0%	0%	10%	10%	10%	10%
Total MCAA Factor	**5%**	**15%**	**25%**	**30%**	**30%**	**30%**
Estimated Loss of Productivity	**76**	**198**	**336**	**554**	**526**	**711**
					Total Hours Lost	**2,401**

In Table 3-6, the MCAA states that the rows indicate the following:

> 1. *The actual work period being measured for impacts.*
> 2. *The actual, payroll craft labor hours (without supervision).*
> 3. *Craft hours deducted for time and material ticket work, the contractor's own deficient work (rework), any estimated, self-inflicted productivity losses, crew hours that the contractor believes were not affected by the changes in scope, such as crews working in areas of the project not proximate to the areas where the changed work was performed, and change order adjustments as described herein.*
> 4. *The resulting "revised actual hours."*
> 5. *The list of the MCAA factor categories being applied.*
> 6. *The resulting estimated loss of productivity for each time period.*
>
> *Note that the total MCAA factor percentage has not been multiplied against the revised craft hours. Instead, the percentages have been totaled, the adjusted hours divided by one plus the decimal percent (i.e., 1.05 for the first period in the table), and that result subtracted from the total, adjusted hours. One significant difference between forward estimated and retroactively estimated productivity loss is that the contractor's actual labor hours already include the loss of productivity. Therefore, it is necessary to calculate the productive hours first to avoid overstating the loss of productivity.*[249]

Thus, the MCAA states:

> *In performing a total project, retroactive loss of productivity calculation, it is necessary to deduct the individual forward priced productivity losses, which may have been included in the*

[249] *Id.* at p. 95.

contractor's individual change orders proposals submitted by the contractor to the owner. This deduction can be included when arriving at the adjusted, actual labor hour total.[250]

3.4.1.4.1.3 Criticisms and Court Decisions

Criticisms of the use of the MCAA factors to estimate productivity loss include the following:

- The arbitrary and subjective nature of the MCAA factors undermines their credibility. A common criticism of an analysis reliant upon MCAA factors is that factors can be added or adjusted as necessary to obtain a pre-determined and beneficial result. The MCAA notes that factors should be applied with care, because the addition of multiple factors can lead to unreliable results.[251]

- A question of objectivity exists because the factors were developed solely by parties with a vested interest, and without the tempering perspective of the other project party. The MCAA manual was developed by a contractor group.[252] The MCAA factors were devised for mechanical contractors, and not for the civil, structural, electrical, instrumentation, or other trades.

- The MCAA factors were developed by MCAA's Management Methods Committee, but are not based on any empirical study determining the specific factors or the percentages of loss associated with the individual factors.[253]

- According to one author, *"no comparable factors exist in the UK or Commonwealth jurisdictions, and there is not any judicial record of their being considered."*[254]

---

250 *Id.* at p. 98.

251 See "Change Orders Productivity Overtime – A Primer for the Construction Industry," MCAA, 2012, p. 88.

252 *Id.* at pp. 102–103. Also see Ibbs, William, "Impact of Change's Timing on Labor Productivity," Journal of Construction Engineering and Management, ASCE 131:11, 2005a, p. 1220.

253 *Id.* at pp. 102–103. Also see Schwartzkopf, William, *Calculating Lost Labor Productivity in Construction Claims,* Second Edition, Aspen Publishers, 2004, Chapter 11.02, Mechanical Contractor's Estimating Manual, p. 204.

254 See Pickavance, Keith, *Delay and Disruption in Construction Contracts,* 3rd ed., T&F Informa (UK) Ltd., London, 2005, pp. 569–572, paragraph 17.118.

U.S. courts and boards have accepted the use of MCAA factors when they are supported by credible expert witness testimony.[255] In *Clark Construction Group*, the board, when determining the contractor's damages under the Jury Verdict method, applied the MCAA factors (even though the contractor did not present these factors), stating, "*We will utilize the productivity factors from the MCAA Manual as the best method to arrive at the percentage estimates of…undeniable productivity losses.*"[256]

Other than cases in the United States, the authors are not aware of any other courts or boards of adjudication that have accepted the MCAA method. However, arbitration panels may have accepted this method for productivity loss estimations, but the detailed basis of the findings of arbitration panels is usually not disclosed. While MCAA factors can be one method used to assess the loss of productivity associated with various impacting events, it should not be the only method, as such estimations are not precise and subject to (often correct) criticism. Unless there is no other alternative, MCAA factors should be utilized in support of, or in conjunction with, a separate analysis to quantify productivity loss claims.

3.4.1.4.1.4 Estimating Cumulative Impact

In addition to estimating the loss of productivity man-hours associated with the impact of direct disruption, the MCAA states that its factors can also be used to estimate the cumulative impact of changes.[257] Although the methods for estimating lost man-hours due to direct disruption and cumulative impact are the same, the difference is in the factors that are selected. The MCAA provides a wide variety of factors, some of which may be applicable to cumulative impact.

The MCAA provides typical causes for each loss of productivity factor and specifically identifies the impact of changes for several factors. For example, the MCAA factor

---

255 See, *e.g.*, *Clark Construction Group, Inc.*, 00-1 BCA ¶ 30,870, VABCA No. 5674 (April 2000); *Clark Concrete Contractors, Inc. v. General Services Administration*, 99-1 BCA ¶ 30280, GSBCA No. 14340 (1999), *Fire Security Systems, Inc.*, 91-2 BCA ¶ 23,743; and *Stroh Corporation,* GSBCA No. 11029, 96-1 BCA ¶28,265. See also Appendices A and B for cases in which the damages were quantified using MCAA Factors.

256 *Clark Construction Group, Inc.*, 00-1 BCA ¶ 30,870, VABCA No. 5674 (April 2000), at p. 42).

257 See "Change Orders, Productivity, Overtime: A Primer for the Construction Industry," MCAA, 2012, p. 83: "*It is clear that a contractor must consider both the direct impacts of a loss of labor productivity caused by a change to the contract scope of work, as well as the cumulative impact of changes in scope to the unchanged work. In the past, many contractors have used the MCAA factors only when 'forward pricing' a loss of productivity component of a change order proposal. In addition to providing updated general instruction on the uses of the MCAA factors, this chapter seeks to explain how the MCAA factors can also be applied equitably and reasonably when retroactively quantifying the cumulative effects of changes on the productivity of a construction project.*"

"*Morale and Attitude*" can be caused by "*multiple contract changes and rework*"; "*Reassignment of Manpower*" can be caused by "*unexpected, excessive changes*"; "*Dilution of Supervision*" can be caused when supervision is diverted to "*analyze and plan change*" or "*stop and replan affected work*"; and there is an MCAA factor for "*Ripple*" that aligns with the theory of cumulative impact.[258]

Proper utilization of the MCAA factors and method (*e.g.*, proper selection of factors and percentages, proper adjustments to "actual man-hours," application of factors to the appropriate man-hours and time frames, etc.) is a potentially viable approach for estimating the cumulative impact of changes. However, as with the use of MCAA factors to estimate direct disruption, one must overcome the criticism that the factors are arbitrary and subjective. This problem is exacerbated in cumulative impact claims, as the owner may (rightfully) argue that arbitrary and subject factors are being used to quantify a theoretical impact.

3.4.1.4.1.5 Impacting the Project Schedule Using the MCAA Factors

The activities in the project schedule may also be impacted with loss of productivity estimates derived from using the MCAA factors. Loss of productivity can increase schedule activity durations unless crews are added to mitigate the additional man-hours required because of the impacting events. The adjusted schedule can affect the project's critical path and time-related costs.

For example, a piping activity that was planned to take 20 work days to complete is impacted by a change that not only adds to the additional scope of work, which increases the piping activity's duration by 5 work days, but also affects the duration by loss of productivity due to factors such as "Dilution of Supervision," "Fatigue," and "Ripple." The additive impact caused by these factors totals 20 percent. Therefore, the 20 percent impact increases the activity duration by an additional 5 work days (0.2 x (20 + 5) = 5 days) due to loss of productivity.

3.4.1.4.2 NAVSEA Guidelines.

In 1982, the U.S. Naval Sea System Command (NAVSEA) issued a document entitled "Guidelines on Factors Influencing Cost for Forward Pricing Change Order Disruption, Delay, Acceleration and Cumulative Effects," often referred to as the NAVSEA guidelines.[259] The NAVSEA guidelines were intended to assist forward pricing of change orders and were utilized in relation to certain new ship construction

258 See "Change Orders Productivity Overtime: A Primer for the Construction Industry," MCAA, 2012, pp. 77–78.

259 See Jozwick, John, *Esq.*, "Arbitrating Shipyard Disputes & Damages," The Arbitrator, Volume 36, No. 4, July 2005. See also, *e.g.*, *Advanced Engineering* & *Planning Corp.*, ASBCA Nos. 53366, 54044, November 2004.

or renovation contracts.[260] Under official Navy policy, the use of the NAVSEA guidelines was entirely discretionary, and even when they were utilized, the factors were subject to negotiation.[261] The authors were unable to confirm whether the NAVSEA guidelines are still used by the Navy.

The guidelines list varying direct disruption factors with respect to the following five categories:

- Number of affected crafts;
- Number of compartments affected;
- Area of disruption, *i.e.*, the congestion in the affected area of work;
- Complexity of the vessel; and
- Time factors.[262]

The extent of disruption increases as the level of complexity increases. For example, the guidelines provide a 2.5 percent loss of productivity factor if between one and five crafts are affected by a change, but a 5 percent loss of productivity factor if between six and 10 crafts are affected.[263]

Notably, the guidelines include separate factors for the cumulative impact of changes to the unchanged work. For example, if a change amounts to a 15 percent increase to contract scope, the guidelines provide for a loss of productivity of between 3.3 percent and 6.7 percent (or 2 to 4 minutes in an hour) to the unchanged work.[264]

While they were intended to assist in the contemporaneous pricing of change orders, the NAVSEA guidelines (or variants) have been used by contractors to retrospectively estimate both local and cumulative disruption, in cases decided as late as 2007.[265] Based upon a review of published cases, courts or boards have not awarded contractors disruption damages when the NAVSEA guidelines were used.

---

260 See *Southwest Marine,* 94-3 BCA ¶ 27,102, DOTCAB No. 1663, 1994 WL 417289 at paragraph 73.

261 *Id.*

262 See Jozwick, John, *Esq.*, "Arbitrating Shipyard Disputes & Damages," The Arbitrator, Volume 36, No. 4, July 2005.

263 *Id.*

264 *Id.*

265 See *Lake Union Drydock Company, Inc. v. U.S.*, US District Court, W.D. Washington, at Seattle, No. C05-2146RSL (October 2007); *Advanced Engineering & Planning Corp.*, ASBCA Nos. 53366, 54044, November 2004; *Triple "A" South,* 94-3 BCA ¶ 27,194, ASBCA No. 46,866 (1994); and *Southwest Marine*, 94-3 BCA ¶ 27,102, DOTCAB No. 1663 (1994).

3.4.1.5 System Dynamics Modeling

The system dynamics (SD) modeling method has been employed to analyze delay and disruption claims. SD computer models on engineering and construction projects attempt to reproduce the performance driving feedback mechanisms, and then use the model to determine how the project's performance would vary under different conditions and/or different management policies. SD computer models rely on feedback mechanisms spanning the engineering, procurement, and construction work to complete a project. Instead of statusing each activity, as in a CPM schedule, these SD models simulate aggregated flows of tasks and the numerous factors that dynamically affect those flow rates. This SD modeling allegedly allows many performance driving factors to be modeled, and enhances the analytical reliability of the analysis of cause-effect in projects that have been affected by delay and disruption.[266]

Proponents of this method contend that SD modeling quantification of cumulative impacts can overcome one of the major limitations of the Measured Mile analysis and productivity comparisons with other projects, which may not properly handle the multiple and/or concurrent disruptions that are caused by different project parties. Users of SD models contend that they can correctly quantify owner-responsible delay and disruption impact costs and demonstrate the cause-effect relationship of the cumulative impact.

As one example, an SD model prepared for a claim on a large engineering and construction project might attempt to duplicate the actual conditions (*e.g.*, man-hours and percent complete) by time for 20 major disciplines of work that encompass the entire project, even though the project schedule includes 5,000 activities. The model would include factors to account for scope changes, owner-caused impacts, and contractor-caused impacts, as well as feedback mechanisms to model project performance for such components as rework and the interrelationships between the

---

266 See a discussion of System Dynamics Modeling in Ibbs, William, Long D. Nguyen, and Seulkee Lee, "Quantified Impacts of Project Change," Journal of Professional Issues in Engineering Education and Practice, ASCE, January 2007, p. 49. Also see Stephen P. Warhoe, PE CCE CFCC and Dr. Paul D. Giammalvo, CCE, "Understanding Change of Scope Effects on DBB Projects Using Systems Thinking," 2011 AACE International Transactions, CDR.501; Moonseo Parka and Feniosky Peña-Morab, "Dynamic change management for construction: introducing the change cycle into model-based project management," System Dynamics Review, Volume 19, Number 3, Fall 2003; Susan Howick, "Using system dynamics to analyse disruption and delay in complex projects for litigation: can the modelling purposes be met?," Journal of the Operational Research Society (2003) 54, pp. 222–229; Stephen P. Warhoe, PE CCE CFCC, Modelling Productivity Loss Using System Dynamics to Understand Cumulative Impact, 2013 AACE International Transactions, PM 1432.

major disciplines of work. These factors and feedback mechanisms attempt to account for delay and man-hour growth from the plan. Once the model is validated to fit actual project performance, the owner-caused impacts are removed to show how the contractor would have performed "but-for" owner-caused impacts, with the claim comprised of the difference between actual and "but-for" man-hours.

While SD models are quite complex, the modeled man-hour growth is still largely driven by user-defined inputs of various impacts, some of which are estimated. An SD analysis must therefore clear the same hurdles as more traditional productivity analysis and demonstrate that, for example, the disruption factors used are reasonable and supportable, and potential contractor-responsible bid error or performance problems are adequately accounted for in distributing responsibility for man-hour and cost overruns.

SD analyses have led to successful out-of-court settlements for the contractor in delay and disruption disputes.[267] However, SD modeling has not achieved as much popularity in construction disputes as other methods for a number of reasons: (i) it is expensive and requires a high level of expertise to prepare;[268] (ii) it requires significant contemporaneous project documentation and data to prepare;[269] and (iii) due to their complexity, the models and results are not readily understood by all litigation audiences.[270]

3.4.2 Industry Studies Associated with Overtime and Overcrowding/Congestion

This section addresses several industry studies related to overtime and overcrowding/congestion, two of the most common labor productivity impacts associated with acceleration. While researchers have developed equations and curves in an attempt to correlate overtime work and overcrowding/congestion with labor productivity loss, the composition and the quality of the underlying data and the analyses used to formulate the results are often unknown. In many cases, the curves are provided without reference or explanation.

---

267 See Lyneis, J. and Ford, D., "System dynamics applied to project management: a survey, assessment, and directions for future research," System Dynamics Review, Volume 23 No. 2/3, 2007, p. 171.

268 See Ibbs, William, Long D. Nguyen, and Seulkee Lee, "Quantified Impacts of Project Change," Journal of Professional Issues in Engineering Education and Practice, ASCE, January 2007, pp. 47, 49.

269 *Id.* at p. 49.

270 *Id.*

Various factors can interact to influence productivity on a construction project. While many industry studies have been performed to determine, or measure, the effects of specific factors upon labor productivity, there has not been significant research to determine how these various factors interact. The studies that have been performed vary as to the amount of empirical data collected and the method that was used to gather the data. In using such industry studies to calculate a result for a given project, one should evaluate the applicability of the underlying data and the purpose for which it is being used.

3.4.2.1 Estimating Loss of Productivity Man-hours Related to Working Overtime

Numerous studies have been published regarding the negative effect on productivity as a result of working overtime. A discussion of the following overtime studies is provided below:

- U.S. Department of Labor Bulletin No. 917;
- Models Published by the Business Roundtable;
- NECA Overtime Study;
- 1988 CII Study;
- 1994 CII Study; and
- U.S. Army Corps of Engineers Study.

3.4.2.1.1 U.S. Department of Labor Bulletin No. 917

The seminal study on the effects of overtime on productivity is Bulletin No. 917, which was published in 1947 by the U.S. Department of Labor.[271] Figure 3-8 presents the data from this study.[272] The data provides a measure of productivity for 5, 6, and 7-day work weeks, and for 8, 9, 10, 11, and 12 hours worked per day.

---

[271] The study compiled productivity data from 34 industrial manufacturing plants, and therefore may not be applicable to construction projects.

[272] See "Hours of Work and Output," United States Department of Labor, Bulletin 917, 1947.

Figure 3-8: U.S. Department of Labor Overtime Productivity Data

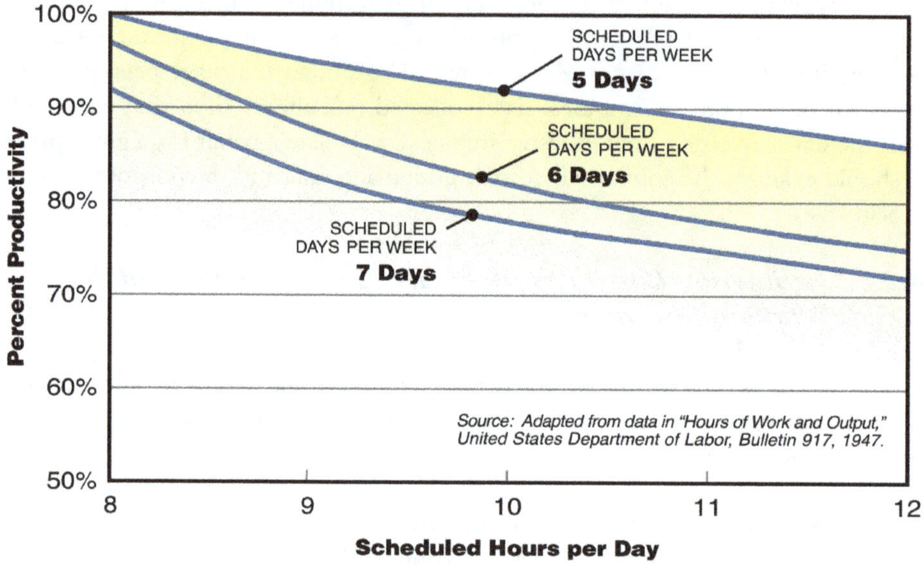

This data shows that working long hours over a consistent period can have a considerably adverse effect on productivity. Thus, although working 8 hours a day for 5 days will produce 100 percent productivity, assuming that there are no other negative impacts, an increase to 12 hours a day over the same period will produce a drop to 86 percent productivity. Working 12 hours a day for 7 days a week will result in a reduction to 72 percent productivity.

Since 1947, a number of further studies have looked at this issue in more detail. These studies are discussed in the following sections.

3.4.2.1.2 Models Published by the Business Roundtable

In 1980, the Business Roundtable published a study specific to the effects of overtime on productivity in construction projects.[273] The study specifically acknowledges that it applies *"only to operations where the total job is placed on an overtime basis for an extended period of time."*[274]

---

[273] See Construction Industry Cost Effectiveness Task Force, "Scheduled Overtime Effect on Construction Projects," Business Roundtable, November 1980. The Business Roundtable Report provides productivity loss data for 50 and 60-hour work weeks, and was based on data from a single factory in Green Bay, Wisconsin, at which a variety of projects were constructed for Proctor & Gamble.

[274] "Scheduled Overtime Effect on Construction Projects," The Business Roundtable, November 1980, p. 2.

The study indicates that there is a significant drop in productivity during the first week of an overtime program, followed by a gradual improvement in productivity that peaks at the third week. From that point onward there is a continuous decline in productivity until the 9th week, at which time productivity losses level out.

Figure 3-9 presents the data from the Business Roundtable study.[275]

Figure 3-9: Effect of Overtime on Productivity – 50 and 60 Hour Workweeks – Business Roundtable Study

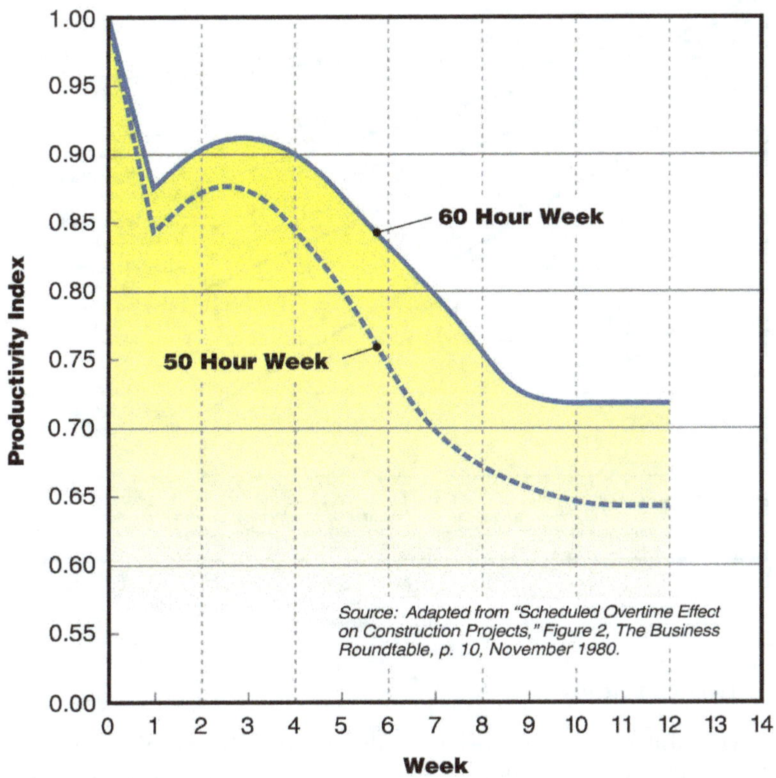

3.4.2.1.3 NECA Overtime Study

The National Electrical Contractors Association (NECA) published data in 1969 for sporadic or isolated overtime, based on a survey of NECA membership rather than empirical data. Therefore, the data may not be as precise as that obtained for the Business Roundtable Report or the U.S. Department of Labor study. However, the NECA study does provide estimates of productivity loss for a much greater range of overtime programs.

---

[275] *Id.* at Figure 2, p. 10.

Figure 3-10 presents the data from the NECA study.[276] Productivity forecasts are presented for 6 and 7-day work weeks and 8, 9, 10, and 12 hours per work day. All of the curves indicate a decline in productivity for the first week of working overtime, and then an additional but lower decline in productivity loss from weeks 2 through 4. Productivity loss was greater for 7 days per week and 10 and 12 hours per day, compared to 7 days per week and 8 and 9 hours per day.

Figure 3-10: Productivity Loss Curves 1969 NECA Study

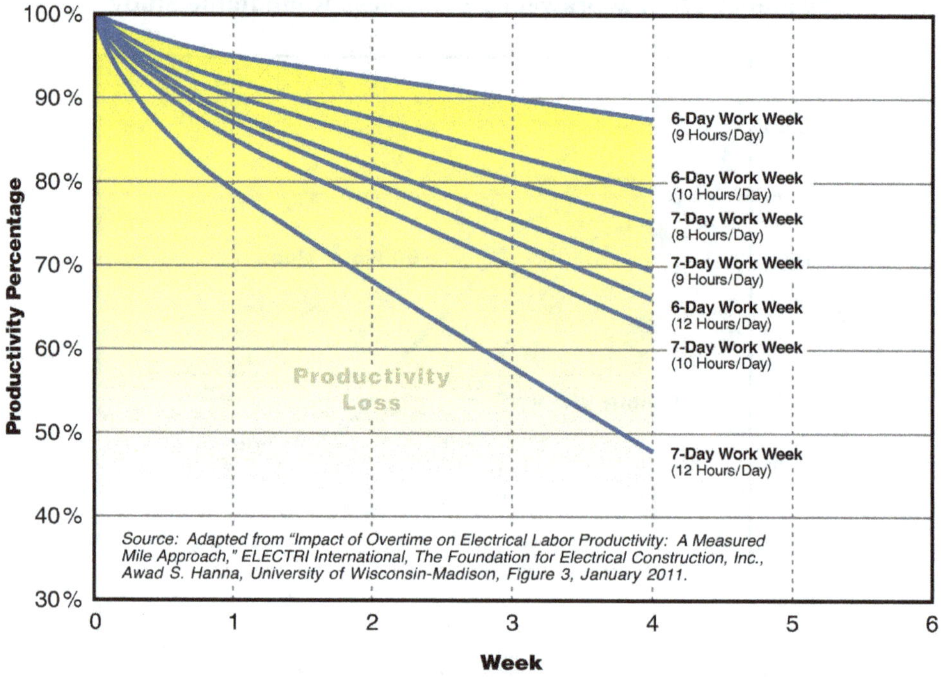

3.4.2.1.4 1988 CII Study

In 1988, CII published an extensive study of the effects of overtime on labor productivity.[277] The data from this study was obtained from seven different industrial plants, including refineries, fossil power plants, chemical plants, and natural gas

[276] See "Impact of Overtime on Electrical Labor Productivity: A Measured Mile Approach," ELECTRI International, The Foundation for Electrical Construction, Inc., Awad S. Hanna, University of Wisconsin-Madison, Figure 3, January 2011 citing data obtained by National Electrical Contractors Association (NECA), Washington , D.C., "Overtime and Productivity in Electrical Construction," 1969.

[277] See "The Effects of Scheduled Overtime and Shift Schedule on Construction Craft Productivity," Construction Industry Institute, Source Document 43, December 1988.

recovery plants. The crafts included insulators, pipe fitters, electricians, carpenters, ironworkers, and laborers.[278]

This CII study indicated that productivity losses from working overtime are not automatic, and it is possible to work 60-hour weeks without serious productivity losses. However, the data generally shows productivity losses from working overtime.

For example, Figures 3-11[279] and 3-12[280] present the productivity data from the CII study regarding large and small bore pipe crews, respectively.

Figure 3-11: Productivity Loss Data – 1988 CII Study
Large Bore Pipe Crews – 4/9s & 1/8

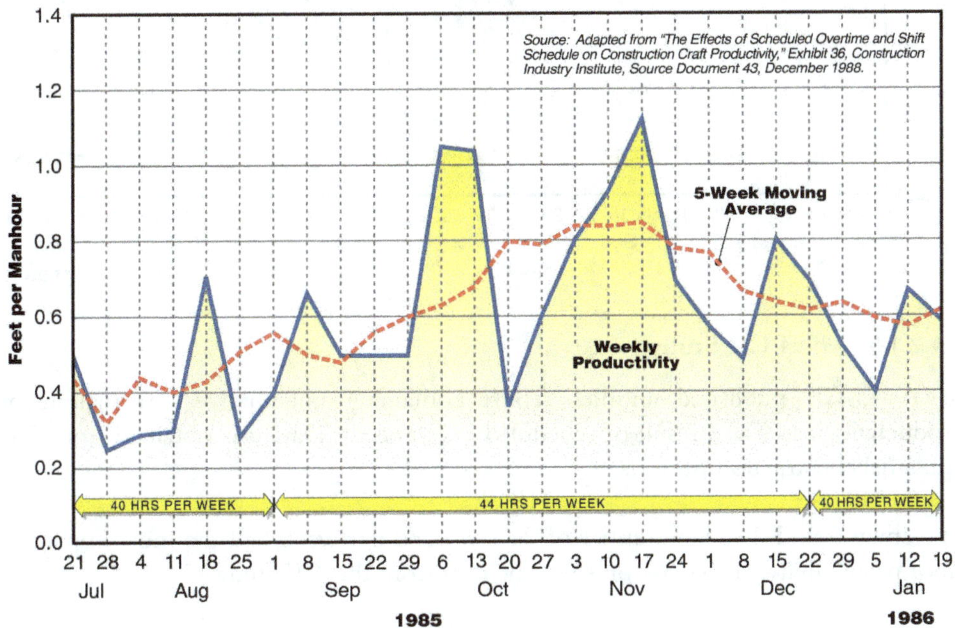

278 See productivity data for these other crafts as reported by the CII Source Document 43 and Source Document 60, November 1990.

279 See "The Effects of Scheduled Overtime and Shift Schedule on Construction Craft Productivity," Exhibit 36, Construction Industry Institute, Source Document 43, December 1988.

280 *Id.* at Exhibit 38.

Figure 3-12: Productivity Loss Data – 1988 CII Study
Small Bore Pipe Crews – 4/9s & 1/8

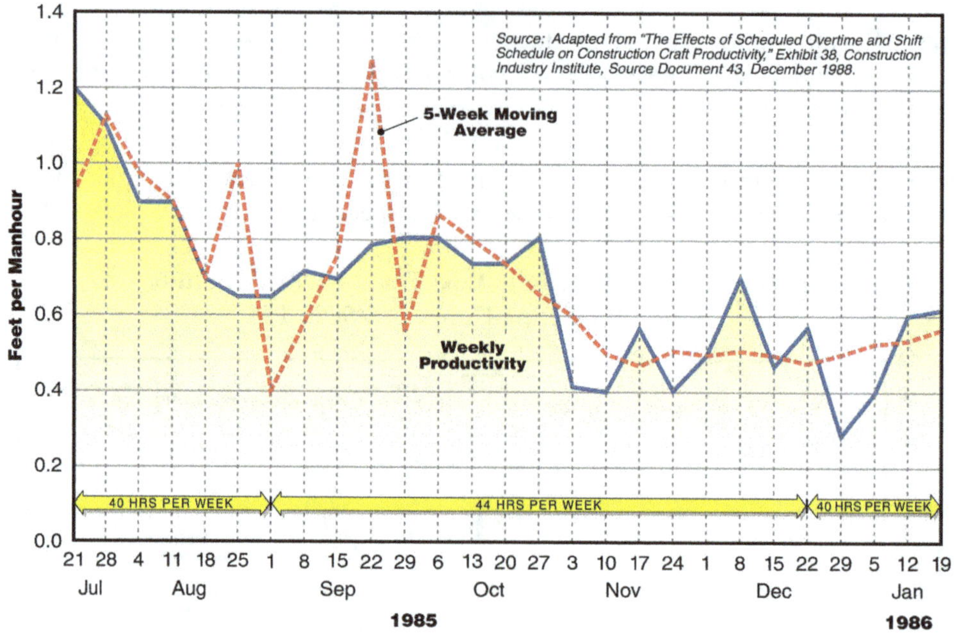

3.4.2.1.5 1994 CII Study

In 1994, CII published another report containing quantitative information on productivity losses as a result of scheduled overtime.[281] Data was obtained from four industrial construction projects.

The study found that the number of days worked per week had the most impact on labor productivity. Table 3-7 presents the data from this CII study.[282]

---

281 See Thomas, H. Randolph and Karl A. Raynar, "Effects of Scheduled Overtime on Labor Productivity: A Quantitative Analysis," Construction Industry Institute, Source Document 98, August 1994. Also see William Schwartzkopf, *Calculating Lost Labor Productivity in Construction Claims,* 2nd Edition, Aspen Publishers, Inc., New York, 2004, §2.06, pp. 29–31.

282 *Id.* Thomas and Raynar, Table 5.

Table 3-7: Labor Overtime Productivity Data – 1994 CII Study

Days Worked Per Week	Number of Data Points	Average Weekly Performance Factor	Efficiency
2	6	0.56	0.57
3	8	0.64	0.65
4	50	0.98	1.00
5	38	0.86	0.88
6	18	0.84	0.86

The authors of this study provide the following definitions:

Weekly Performance Factor = Baseline Productivity/Weekly Productivity

Efficiency = Average Weekly Performance Factor/Average Weekly Performance Factor for a 40-hour 4-day work week or 0.98. Thus, the baseline productivity of 100 percent (1.0 efficiency) was established for a 4-day, 10-hour per day work week. For a 2-day work week, the efficiency would be (0.56/0.98) = 0.57 or 57 percent productivity. For a 6-day work week, the efficiency would be (0.84/0.98) = 0.86 or 86 percent productivity.

3.4.2.1.6 U.S. Army Corps of Engineers Study

In July 1979, the U.S. Army Corps of Engineers (COE) published the Modification Impact Evaluation Guide,[283] since rescinded, which contained a chart showing the effects of work schedules on efficiency. The curves produced in this document only span a 4-week period, but provide a basis for estimating efficiency loss for 5, 6 and 7-day weeks and 8, 9, 10, and 12-hour days.

Figure 3-13 presents the data from the COE study.[284]

283 See "Modification Impact Evaluation Guide," U.S. Army Corps of Engineers, Department of the Army, EP 415-1-3, July 1979. The Corps of Engineers has rescinded this publication.

284 *Id.* at Figure 4-4, p. 4-17, July 1979.

Figure 3-13: Productivity Loss Data – 1979 U.S. Corps of Engineers

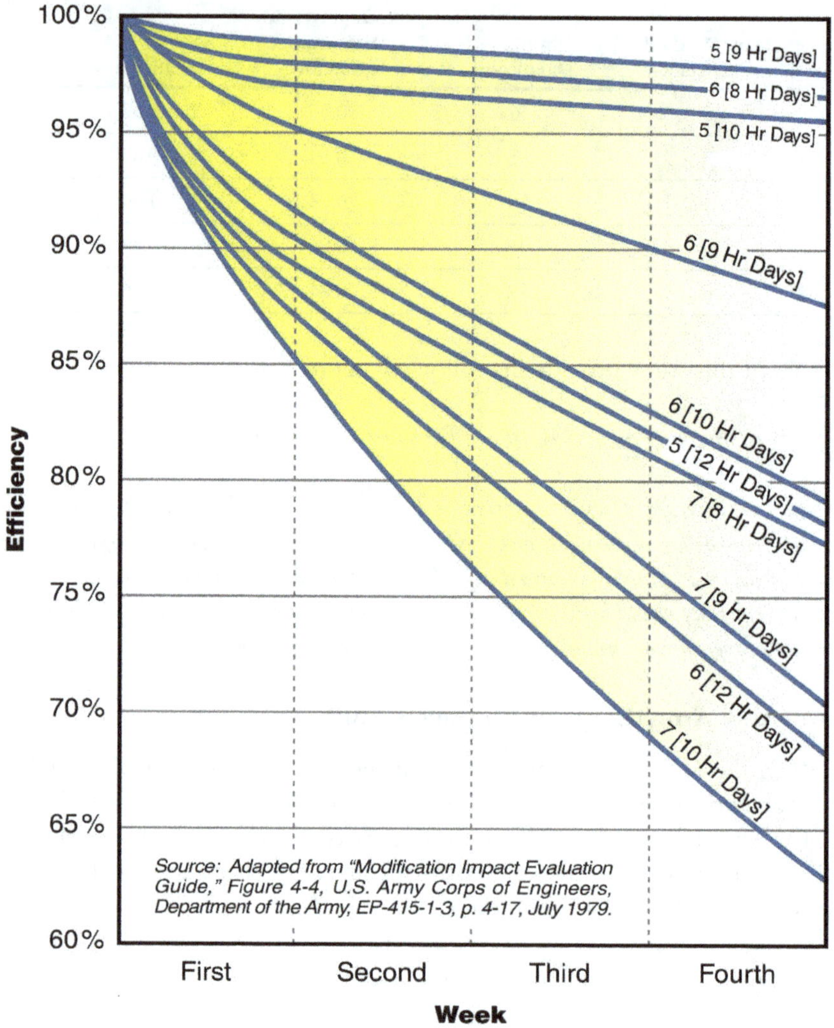

Using these curves, productivity loss is predicted to be less than 5 percent for 5-day work weeks and either 9 or 10 hours per day worked, as well as for a 6-day work week but only 8 hours per day worked. However, these curves predict a much greater productivity loss for 6 and 7-day work weeks with 9, 10, or 12 hours worked per day. Only 65 percent productivity would be predicted when working seven, 10-hour days, with additional declines in productivity for each additional week worked.

3.4.2.2 Estimating Loss of Productivity Man-hours Related to Overcrowding/Congestion

Overcrowding occurs when more workers are placed in a given area than can function effectively. The terms crowding, congestion, overmanning, increased worker density, and trade stacking are often used synonymously. Several industry studies have attempted to quantify the resulting loss of productivity associated with overcrowding. The studies are separated into two groups: those using worker densities as a basis, and those using comparisons between planned and/or actual manpower as a basis.

3.4.2.2.1 Studies Based on Worker Density

Several studies have correlated labor productivity with worker density, or the average square footage that each craftsman has to work. In a 1990 paper, H. Randolph Thomas and Gary Smith outlined three such studies, the results of which are shown in Figure 3-14.

Figure 3-14: Lost Efficiency Based on Worker Density

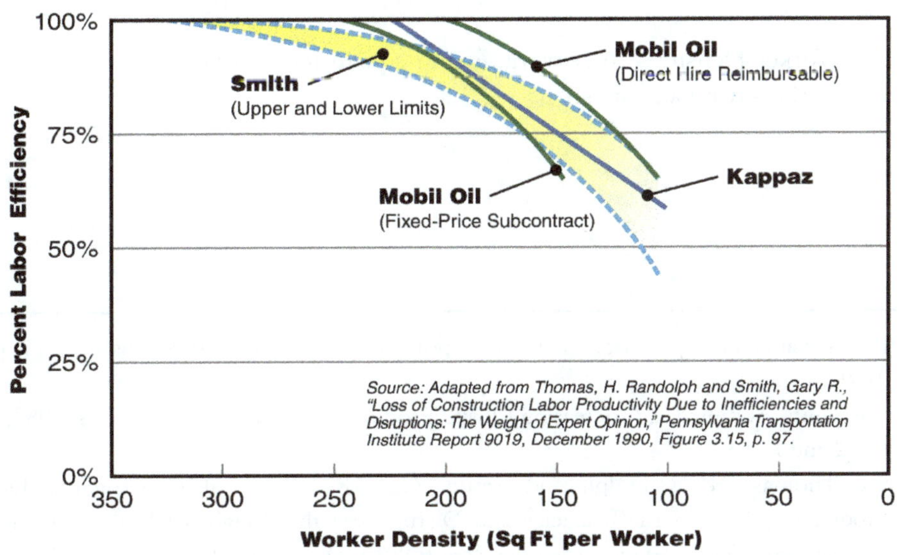

All three studies indicate that labor productivity worsens as a craftsman's average work area decreases. The unpublished Mobil Oil Study relates to refinery construction and indicates that productivity decreases when the average work area falls below 250 square feet per worker for fixed price subcontract labor, and below 200 square feet per worker for direct-hire reimbursable labor.[285] Smith provided a

[285] See Thomas, H. Randolph and Smith, Gary R., "Loss of Construction Labor Productivity Due to Inefficiencies and Disruptions: the Weight of Expert Opinion,"

chart showing a range of productivity loss as worker density increased, with productivity loss starting when the average work area falls below 323 square feet (30 square meters) per worker.[286] However, the chart was provided without reference or explanation as to its source. According to Thomas and Smith, Kappaz' results are based on an unpublished work, and no details were provided with respect to the 16 data points used in his analysis.[287]

More recently, Dr. Awad Hanna provided results from a separate study in a document entitled "Factors Affecting Productivity for Electrical Contractors updated 2007." The study addressed, in part, stacking of trades, which Hanna described as the total number of craftsmen from all trades working in a given area.[288] The study was based on data collected from 152 projects that were competitively bid under design-bid-build contracts. Project sizes ranged from 2,000 man-hours to 150,000 man-hours, and included commercial, institutional, and industrial projects.[289] Hanna did not provide details of the method used to establish his results. He provided the following equation and plot in Figure 3-15:

> Loss of Efficiency = 0.938 – 0.00492*(Worker Density), where:[290]
>
> Worker Density = (net work area in square feet)/(total number of workers from all trades)

Pennsylvania Transportation Institute Report 9019, December 1990, Figure 3.15, pp. 91–92.

286 See Smith, Andrew G., "Increasing Onsite Production," AACE Transactions, 1987, pp. K.4.2 and K.4.3.

287 See Thomas, H. Randolph and Smith, Gary R., "Loss of Construction Labor Productivity Due to Inefficiencies and Disruptions: the Weight of Expert Opinion," Pennsylvania Transportation Institute Report 9019, December 1990, Figure 3.15, p. 93.

288 See Hanna, Awad, "Factors Affecting Productivity for Electrical Contractors updated 2007," ELECTRI International, The Foundation for Electrical Construction, Inc., October 2007, p. 17.

289 *Id.* at p. 5.

290 Hanna, Awad, "Factors Affecting Productivity for Electrical Contractors updated 2007," ELECTRI International, The Foundation for Electrical Construction, Inc., October 2007, p. 18. The actual equation provided in the paper [Loss of Efficiency = 0.938 + 0.00492*(Worker Density)] does not match Hanna's plot of the Stacking of Trades equation or Hanna's example, and is seemingly in error. The equation shown above using "–" instead of "+" does match the plot and is assumed to be correct.

Figure 3-15: Lost Efficiency Due to Trade Stacking (Hanna)

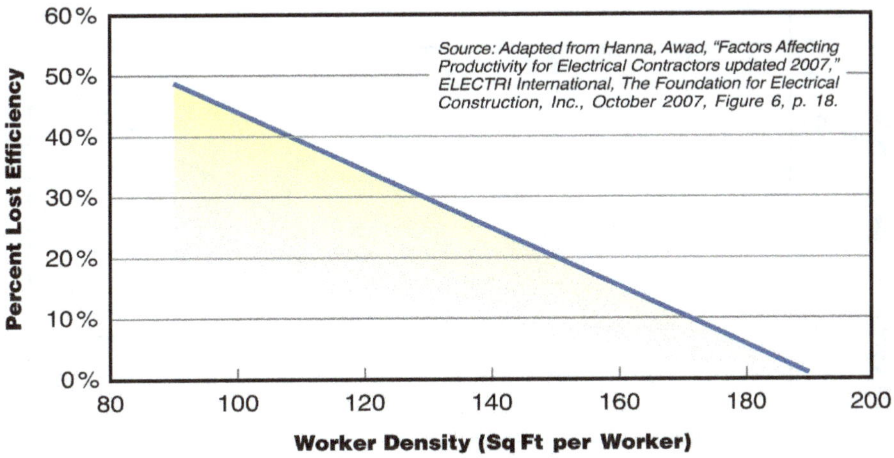

Hanna's results suggest that a linear relationship exists between worker density and labor productivity, with craftsmen becoming inefficient when their work area drops below approximately 190 square feet on average, as shown in Figure 3-15. Hanna notes that the loss of efficiency percentage should only be applied to those man-hours that were worked in the area during the overcrowded conditions, and that the above equation is only applicable for projects which range between 2,000 and 150,000 total man-hours, and worker densities which range between 90 and 191 square feet per worker.[291]

3.4.2.2.2 Studies Comparing Planned and/or Actual Manpower

In July 1979, the U.S. Army Corps of Engineers (COE) published the Modification Impact Evaluation Guide,[292] since rescinded, which included the below curve in Figure 3-16 on the effect of crowding. At 15 percent overcrowding, for example, the curve predicts an approximate 5 percent loss of efficiency.

The guide provided no information regarding the source of this curve. Furthermore, it cautioned that one cannot assume that a contractor's planned or scheduled manpower is the proper basis for establishing the percent crowding, as it is possible that more workers can work efficiently than the contactor had planned. As a result, the curve should be interpreted as representing the upper limit of productivity loss.[293]

---

291 *Id.* at p. 18.

292 See "Modification Impact Evaluation Guide," U.S. Army Corps of Engineers, Department of the Army, EP 415-1-3, July 1979. The Corps of Engineers has rescinded this publication.

293 *Id.* at pp. 4–8.

Figure 3-16: Effect of Crowding on Labor Efficiency

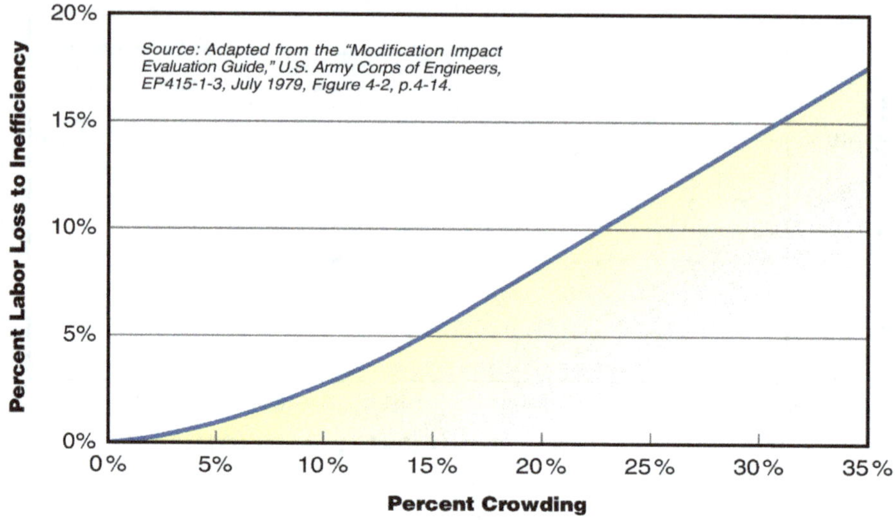

In a 1990 paper, H. Randolph Thomas and Gary Smith provided four curves related to overmanning, the results of which are shown in Figure 3-17.

The curves are based on a planned workforce of 600 craftsmen and sourced from a technical paper by O'Connor, a textbook by Waldron, and two unpublished sources developed by general contractors.[294] Thomas also noted a problem with using these studies: defining the correct base from which overmanning is measured.[295] If contractors plan to use these studies, they must demonstrate that their planned manpower loading, before the impacts were incurred, was based on an optimum number of workers (for the given area). The planned man-loading per area of workspace could have been well below the optimum. Thus, an increase to planned manpower does not necessarily cause overcrowding and a loss of productivity.

Dr. Awad Hanna has prepared two studies related to overmanning, a term which he defines as "*…an increase of the peak number of workers of the same trade over and above the planned or estimated peak.*"[296] These studies differ from Hanna's study of stacking of trades (described in Section 3.4.2.2.1 above) in that they address the congestion associated with one trade, rather than all trades. Neither of Hanna's studies actually

---

294 See Thomas, H. Randolph and Smith, Gary R., "Loss of Construction Labor Productivity Due to Inefficiencies and Disruptions: the Weight of Expert Opinion," Pennsylvania Transportation Institute Report 9019, December 1990, p. 140.

295 *Id.*

296 Hanna, Awad, "Factors Affecting Productivity for Electrical Contractors updated 2007," ELECTRI International, The Foundation for Electrical Construction, Inc., October 2007, p. 7.

uses planned manpower as a basis for comparison to actual manpower; rather, Hanna uses the ratio of actual peak manpower to actual average manpower.

Figure 3-17: Comparison of Overmanning Curves for a Base Work Force of 600 Craftsmen

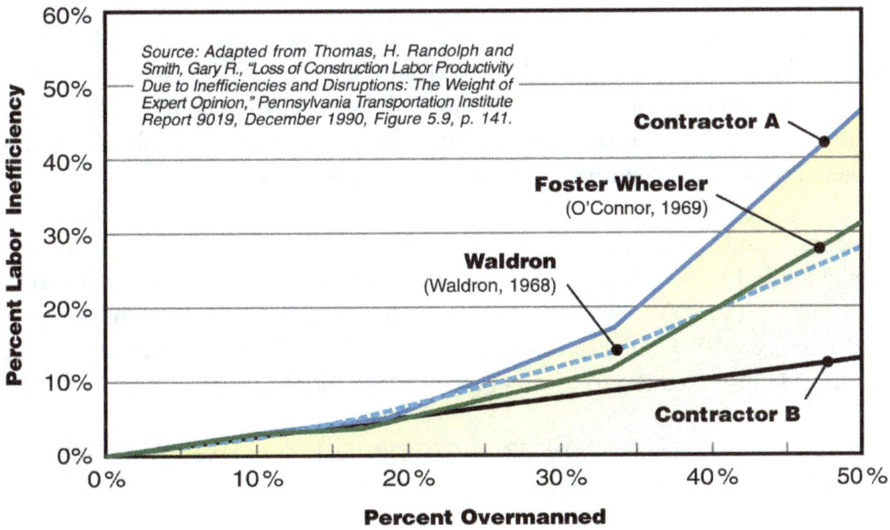

A contractor's actual peak manpower will exceed its average actual manpower on most typical projects. Hanna reviewed prior industry studies and found that the ratio (peak manpower/average manpower) typically falls between 1.35 and 1.6, and he therefore considers overmanning to have occurred when the peak to average ratio exceeds 1.6.[297] This is applicable in the two studies described below.

In 2005, Hanna and others published a paper for the Construction Research Congress using data collected from mechanical and sheet metal contractors.[298] The team collected data from 104 projects, 54 of which met both criteria of having labor productivity loss and having a peak actual manpower to average actual manpower ratio larger than 1.6.[299] Based upon multiple regression analysis of the data, the team determined the following formula:[300]

297 See Hanna, A.S., Chang, C., Lackney, J.A., and Sullivan, K.T., "Overmanning Impact on Construction Labor Productivity," Construction Research Congress, American Society of Civil Engineers (ASCE), 2005, p. 6.

298 See Hanna, A.S., Chang, C., Lackney, J.A., and Sullivan, K.T., "Overmanning Impact on Construction Labor Productivity," Construction Research Congress, American Society of Civil Engineers (ASCE), 2005, p. 1.

299 *Id.* at 5.

300 *Id.* at 6.

> Loss of Efficiency = – 0.305 + 0.116*(Act. Peak/Avg) + 0.163* Log (Act. Peak), where:
>
> Act. Peak/Avg = (Actual Peak Manpower)/(Actual Average Manpower), and Act. Peak = Actual Manpower at Peak

The team noted that the formula is only applicable for projects that fall within the bounds of the data collected, which is: project size between 700 and 208,451 man-hours; peak to average ratio between 1.7 and 3.76; and actual peak manpower between 4 and 50 workers.[301] The quantified loss of efficiency is only applicable to portions of the project impacted by overmanning, and not the entire project.[302]

The second study was based on data collected from 152 projects that were competitively bid under design-bid-build contracts. Project sizes ranged from 2,000 man-hours to 150,000 man-hours, and included commercial, institutional, and industrial projects.[303] Hanna did not provide details of the method used to establish his results. He provided the following equation and plot in Figure 3-18:

> Loss of Efficiency = 0.0733 + 0.0741*(Peak over Average Ratio), where:[304]
>
> Peak over Average Ratio = (Peak Number of Craftsmen)/(Average Number of Craftsmen)

Hanna noted that the formula is only applicable for projects that fall within the bounds of the data collected, i.e., a project size between 2,000 and 150,000 total man-hours and a Peak over Average Ratio between 1.70 and 3.95.[305] Again, the quantified loss of efficiency is only applicable to those portions of the project impacted by overmanning, and not the entire project.[306]

---

301 *Id.*

302 *Id.* at 8.

303 See Hanna, Awad, "Factors Affecting Productivity for Electrical Contractors updated 2007," ELECTRI International, The Foundation for Electrical Construction, Inc., October 2007, p. 5.

304 Hanna, Awad, "Factors Affecting Productivity for Electrical Contractors updated 2007," ELECTRI International, The Foundation for Electrical Construction, Inc., October 2007, p. 9. The actual equation provided in the paper [Loss of Efficiency = 0.<u>00</u>733 + 0.0741(Peak over Average Ratio)] does not match Hanna's plot of the overmanning equation, and is seemingly in error. The equation shown above using 0.<u>0</u>733 does match the plot and is assumed to be correct.

305 *Id.* at p. 9.

306 *Id.*

Figure 3-18: Lost Efficiency Due to Overmanning (Hanna)

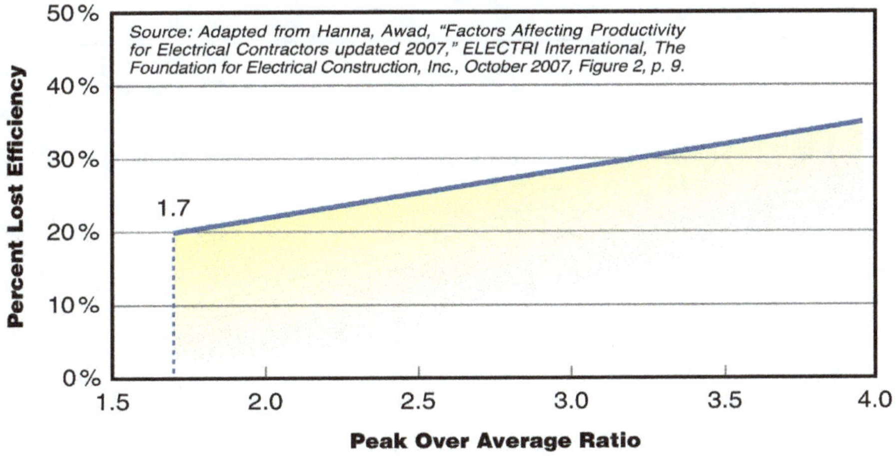

Chapter 4

Example Calculations for Estimating Productivity Loss

Chapter 4

Example Calculations for Estimating Productivity Loss

Chapter 4 provides example calculations from the industry studies that were previously discussed. In Section 4.1, example calculations using the Leonard Study models, the two CII studies, the MCAA Factors, and the 2005 Ibbs Study are presented. In Section 4.2, the results from using each of these industry studies are compared.

The calculations will be based upon Table 4-1, showing data in direct construction man-hours. For this example, assume that the work was mechanical in nature and that the contractor accelerated its work to meet its contractual completion date. Further, from detailed cost records, contractor-caused errors and other non-compensable time are estimated to total approximately 10,000 man-hours.

Table 4-1: Man-Hour Data – Mechanical Project

Source of Man-hours	Man-Hours
Base Contract, Earned	120,000
Approved COs, Earned	30,000
Revised Contract, Earned	150,000
Actual MHs	200,000
MH Variance	50,000
Contractor Performance Problems	10,000

4.1 Example Calculations from Industry Studies

Calculations of the estimated productivity loss resulting from the cumulative impact of change using the following industry studies are discussed below:

- 1998 Leonard Study Models (see Section 4.1.1);
- 1995 CII Study (see Section 4.1.2);
- 2001 CII Study (see Section 4.1.3);
- MCAA Factors; (see Section 4.1.4); and
- 2005 Ibbs Study (see Section 4.1.5).

The quantification of percent change and lost man-hours can differ between studies and, therefore, care must be taken to follow the procedures set out in the studies themselves.

4.1.1 1998 Leonard Study

Leonard provides several simple steps to estimate productivity losses:[307]

1. Determine the actual change order man-hours or, in absence of actual man-hours, use earned man-hours. In this example, the value is 30,000 earned change order man-hours.

2. Calculate the "actual contract hours," which is the total actual man-hours, less the earned change order man-hours, less man-hours attributable to contractor-caused performance problems. The value in this example is 160,000 actual contract hours (200,000 – 30,000 – 10,000).

3. Calculate the percent change order by dividing the earned change order man-hours by the actual contract hours. The result is 0.1875, or 18.75 percent (30,000/160,000).

4. Use the appropriate model to determine percentage loss of productivity based upon percent change orders. In this example, the model for mechanical and electrical work would be used.[308] Further, because the example project was also impacted by acceleration, the line including changes plus one major cause should be used. From the table, a 28 percent productivity loss is calculated.

5. Calculate the unproductive hours on the original contract work by multiplying the percent loss of productivity by the actual contract man-hours. The result is 44,800 unproductive man-hours (28 percent x 160,000).

The 44,800 man-hours calculated are more than would be claimed using a Modified Total Cost approach, where the contractor would claim only 40,000 man-hours. Note, however, that if the acceleration was not determined to be a "major cause," the revised productivity loss would be approximately 17 percent, yielding 27,200 unproductive man-hours, and the 17,600 additional man-hours may be attributable to

307 See Charles A. Leonard, "The Effects of Change Orders on Productivity," M.S. Thesis at Concordia University, 1988, pp. 125–126.

308 See Figure 3-2 in Chapter 3, Section 3.4.1.3.1 herein.

acceleration. Due to the general nature of the model, the answer is likely somewhere in between.

4.1.2 1995 CII Study

CII Publication 43-2, entitled "Quantitative Effects of Project Change," provides several models for estimating the cumulative impact of changes.[309] For this example, the most applicable model is that comparing the Construction Change % to the Construction Productivity Ratio.[310] The steps provided by CII are as follows:[311]

1. Calculate the change percentage by dividing the change order man-hours by the total revised budget man-hours (which includes change order man-hours). This value is 20 percent (30,000/ (120,000 + 30,000)).

2. Use the appropriate model to determine the productivity ratio based upon change percentage.[312] Using the Construction Change % to the Construction Productivity Ration curve, the productivity ratio in this example is approximately 95 percent.

3. Calculate the expected final man-hours by dividing the revised budgeted man-hours by the productivity ratio. This results in 157,895 expected final man-hours (150,000/0.95).

4. Calculate the expected man-hours if no changes occurred. As discussed in Chapter 3, Section 3.4.1.3.3, this adjustment is likely not appropriate for quantifying claimed man-hours, because it would result in the contractor claiming for better than planned productivity. In this example, this step has been excluded.

5. Calculate the lost man-hours due to change orders by taking the difference between the expected final man-hours with changes and the earned man-hours. In this example, the calculation shows 7,895 hours lost (157,895 – 30,000 – 120,000).

---

309 See Project Change Management Research Team, "Quantitative Effects of Project Change," CII Publication 43-2, (May 1995), using data from Ibbs, C.W. & Walter E. Allen, "Quantitative Impacts of Project Change," CII Source Document 108 (May 1995).

310 *Id.* at p. 8.

311 *Id.*

312 See Figure 3-4 in Chapter 3, Section 3.4.1.3.3 herein.

Using this study, the contractor would claim 7,895 man-hours for the cumulative impact of changes. The impact of acceleration is not included in the results.

4.1.3 2001 CII Study

CII Research Report 158-11 entitled, "Quantifying the Cumulative Impact of Change Orders for Electrical and Mechanical Contractors" provides the following equation to predict the impact of change orders on productivity:[313]

%Delta = 0.36866 + 0.11957 Percent Change – 0.08065 PM%TimeOnProject – 0.16723 %OwnerInitiatedCO – 0.09147 Productivity – 0.05213 Overmanning + 0.022345 Processing Time

The following values are calculated based upon the project information, some of which were assumed for this example:

- Percent Change: (CO MHs/Base Contracts MHs) = (30,000/120,000) = 0.25
- PM%TimeOnProject: (Assumed 50 percent) = 0.5
- %OwnerInitiatedCO: (Assumed 50 percent) = 0.5
- Productivity: (Contractor did track productivity) = 1.0
- Overmanning: (Overmanning occurred) = 1.0
- Processing Time: (Assumed 14 days for average CO approval) = 2[314]

Inputting these values into the above equation, the final %Delta value is calculated as 0.1757, or 17.57 percent. This value is multiplied by actual hours incurred: (17.57%)(200,000 mhs) = 35,140 mhs. This corresponds to 35,140 man-hours that could be attributed to changes and their effect on the contractor's labor efficiency.[315] However, this value includes all inefficient work hours, including those caused by the contractor. After deducting 10,000 man-hours in contractor-caused losses, the contractor's claim for cumulative impact of changes would be 25,140 man-hours.

---

313 See Awad S. Hanna, "Quantifying the Cumulative Impact of Change Orders for Electrical and Mechanical Contractors," CII Research Report 158-11, at 108 (February 2001).

314 *Id.* at p. 109 (Table 6.2 – Factors Defined).

315 *Id.* at pp. 119–120 (provides example calculation).

4.1.4 MCAA Factors

In determining their models, the Leonard Study and the CII studies collected data on a total project basis and, therefore, the analysis of loss of productivity using these models must be calculated on a total project basis. Using MCAA factors, the contractor is not confined to determining the losses on a total project basis; the impacted scope of work can be segregated so that the correct loss of productivity factors can be applied. Such is the case in this hypothetical example.

A qualified expert has reviewed the project and cost records (including Table 4-1) and has interviewed several project participants. The expert has determined that the contractor suffered productivity losses due to the cumulative impact of changes and acceleration. Using the MCAA factors, the expert has determined the following productivity losses:[316]

- Cumulative impact of changes: a 5 percent productivity loss due to minor "Morale and Attitude" impacts; a 5 percent loss for minor "Reassignment of Manpower" impacts; a 10 percent loss for minor "Dilution of Supervision" impacts; and an additional 10 percent productivity loss due to minor "Ripple" impacts. The total productivity loss calculated for the cumulative impact of change is 30 percent.

- Acceleration: a 20 percent productivity loss due to average "Stacking of Trades" impacts; a 10 percent loss due to minor "Crew Size Inefficiency" impacts; and a 10 percent loss due to minor "Overtime" impacts. The total productivity loss calculated for acceleration is 40 percent.

However, the contractor did not suffer this loss of productivity from day one. The expert determined that the excessive change orders did not start impacting the project until it was approximately one-third complete. Also, the contractor accelerated during the last quarter of the project only. Using base scope man-hours, the expert calculated the lost man-hours as follows:

Cumulative impact of changes:

(120,000 base scope mhs) (66% of mhs impacted) (30% loss of productivity) = **24,000 lost mhs**

316 See "Change Orders Productivity Overtime: A Primer for the Construction Industry," MCAA, 2012, pp. 77–78.

Acceleration:

(120,000 base scope mhs) (25% of mhs impacted) (40% loss of productivity) = **12,000 lost mhs**

Based upon the use of MCAA factors, the contractor might claim the costs for 36,000 man-hours of loss of productivity caused by the cumulative impact of change orders and acceleration.

4.1.5 2005 Ibbs Study

Based on Ibbs' chart and the formulae that are based on using his 2005 project data discussed above in Chapter 3, Section 3.4.1.3.5, one can estimate the Productivity Index for the example project.

The first formula is stated below:

Percent change orders = change order labor hours /(total actual labor-hours – change order hours – contractor mistakes)

Thus, the percent change orders is calculated as follows:

Percent change orders = 30,000 /(200,000 – 30,000 – 10,000) = 30,000 /160,000 = 0.1875 or 18.75%

Using this definition, Ibbs' formula can be used to correlate "% Change Orders" to "Construction Productivity Index" or PI, where the Loss of Productivity (LOP) is defined as follows:

LOP = (1 – PI) x 100%, where PI = earned hours/actual hours

For this example, if the Percent Change Orders was 18.75 percent (x = 0.1875), using Ibbs' formula, one would derive the following estimate of loss of productivity:

$$PI = 2.4621x^2 - 2.169x + 1.0589$$

$$PI = 2.4621 \times (0.1875)^2 - 2.169\,(0.1875) + 1.0589$$

$$PI = 0.0866 - 0.4067 + 1.0589$$

$$PI = 0.7388$$

LOP = (1 – 0.7388) x 100% = 26.12, or approximately 26 % of the actual hours = 52,000 hours

4.2 Comparison of Results

The results of the five analyses are shown in Table 4-2.

Table 4-2: Results from Industry Studies

Industry Study	Lost Man-Hours Due to Changes	Lost Man-Hours Due to Acceleration	Total
Leonard Study	27,200	17,600	44,800
1995 CII Study	7,895	N/A	7,895
2001 CII Study	25,140	N/A	25,140
MCAA Factors	24,000	12,000	36,000
2005 Ibbs Study	52,000	N/A	52,000
Average	**27,247**	**14,800**	**33,167**

The calculations of loss of productivity due to the cumulative impact of changes are somewhat inconsistent, showing a low value of 7,895 man-hours and a high value of 52,000 man-hours. This discrepancy is not unexpected. Industry studies are models used to estimate loss of productivity, and all of these studies warn generally that results may vary and that productivity losses are project specific. Therefore, the losses that are experienced on any given project being analyzed may be different from the projects that are used for these studies.

However, these quantifications would be helpful in supporting a contractor's claim. Recall that the contractor, in this example, expended 50,000 man-hours above its revised budget, of which 10,000 man-hours were estimated to be attributable to the contractor. Using a Modified Total Cost method alone, the contractor would claim 40,000 man-hours. Assuming that the contractor had insufficient data to employ a Measured Mile or Earned Value analysis, the above quantifications generally support the magnitude of such a claim: the average of the five quantifications above is 33,167 man-hours; and excluding the low value derived from the 1995 CII Study, the average increases to 39,485 man-hours.

The presentation of the contractor's actual losses (which, in the authors' experience, is often not provided but should be), coupled with a quantification of lost man-hours from industry studies, is likely the best attempt at demonstrating damages under this scenario. Moreover, the use of multiple quantifications can work to circumvent criticism of one particular study.

Chapter 5

Choosing a Quantum/Damages Method

Chapter 5

Choosing a Quantum/Damages Method

Calculating and proving damages on a construction claim is a difficult undertaking, but one that is made easier by understanding certain fundamental claim principles, including the legal rights of the parties, the general rules of damages, and the burden of proof.

On a construction project, a request or claim for additional compensation is commonly initiated by the contractor, subcontractor, design engineer, or construction manager. Although less common on a construction project, the owner may, under certain circumstances, also be able to recover damages from these above-mentioned parties.[317] Regardless of who is the initiator, a claim must contain three essential elements before any damages can be recovered:

- Entitlement. The actions and/or inactions by the other party or parties to the contract that resulted in compensable liabilities.

- Cause and Effect. Linking a direct relationship between the entitlements for the problems encountered to the financial damages that were incurred.

- Quantum/Damage computation. The method(s) that will be used for presenting and pricing the compensable (recoverable) damages.

There are numerous entitlements that are recognized by U.S. federal courts, state courts, and arbitration panels leading to the recovery of damages related to a construction claim. These entitlements may also be applicable in international dispute resolution forums. These entitlements include: (1) Acceleration, (2) Acts of God/Weather, (3) Cardinal change (breach of contract), (4) Constructive change, (5) Defective & deficient contract documents, (6) Delays, (7) Differing site conditions, (8) Directed change, (9) Disruption, (10) Implied warranty, (11) Impossibility of performance, (12) Mal-administration, (13) Owner-furnished items, (14) Strikes, (15) Superior knowledge, (16) Suspension, (17) Termination, (18) Unjust enrichment, and (19) Variations in quantities.

---

317 Privity of contract would limit the owner from assessing damages directly against the subcontractor; therefore, the owner would have to assess damages against the prime contractor, who would then assess damages against the subcontractor.

Cause and effect relationships linking the entitlements to the damages are most often found in the project's contemporaneous documentation, *i.e.*, meeting records, daily logs, schedules, letters, memos, diaries, photographs, inspection reports, requests for information, monthly progress reports, daily construction reports, etc. Damages, however, depend on detailed and verifiable man-hour and cost data.[318] Having demonstrated the nature of the loss and the connection between the loss and the other party's actions and/or inactions, the success of the claim will often hinge on how well the claimant can calculate and prove its damages. Conversely, the parties to a contract should understand the principles of proof of damages to properly and fairly evaluate a claim, grant the request if it is justified, or reject the claim if convincing and supportable evidence is not presented.

In a civil action for breach of contract, the objective of the trier of fact[319] is to place the claimant as far as possible in the same financial position it would have been in had the alleged breach not occurred.[320] All that is necessary to charge the defendant with the particular loss is that the loss be one that ordinarily flows from the breach of such a contract in the usual course of events, or that reasonable people in the position of the parties would have foreseen as a probable result of the breach.[321] If damages are determined to be direct, they are compensable. If damages are determined to be consequential, they are compensable only if it is determined that the special circumstances were within the contemplation of both contracting parties, and not excluded by the contract.[322]

A contractor pursuing a claim has the burden of proving every element of its case, including the damages that flow from the alleged problem.[323] In an action for damages, the contractor bears the burden of proving both the existence and the amount of the damages incurred.[324] The burden is on the contractor to establish the proximate cause between the breach and the damages, and if the loss caused by the breach cannot be isolated from that which is attributable to other factors, damages may not be available.[325] If the contractor can show: (i) entitlement to recover for the owner's wrongful conduct; (ii) a linkage between the wrongful conduct and the effect; and (iii) damage incurred as a result of that wrongful conduct, the contractor may recover, even though the amount of the damage is uncertain or is based on estimates.

---

318 See Chapter 6, Sections 6.4 and 6.5, and Appendix C herein.

319 The trier of fact can be the jury, judge, board, arbitration panel, or third party decision-maker.

320 See *Maxwell v. Schaefer,* 112 A.2d 69 (1955).

321 See *R. J. Lampus Co. v. Neville Cement Products Corporation,* 378 A.2d 288 (1977).

322 See *Roanoke Hospital Association v. Doyle & Russell, Inc.,* 214 S.E.1d 155 (1975).

323 See *Rochey Bros., Inc. v. Rhoades,* 527 F.2d 891 (1975).

324 See *U.S. v. J. H. Copeland & Sons,* 568 F.2d 1159 (1928).

325 See *S. J. Groves & Sons Co. v. Warner Co.,* 576 F.2d 524 (1928).

As discussed in Chapter 2, Section 2.2.3, many cumulative impact claims fail in court because the calculation of damages is flawed. Yet courts do not require a dollar-by-dollar accounting of the contractor's claim. A claimant need not prove his damages with absolute certainty or mathematical exactitude. It is sufficient if the claimant furnishes the court with a reasonable basis for computation, even though the result is only approximate.[326]

In Section 5.1 below, the concepts of cost variance analysis and man-hour variance analysis are discussed.

5.1 Preliminary Analyses

Preparing a cost variance analysis (CVA) and man-hour variance analysis (MVA) are the first steps in any quantum calculation from which a disruption and/or cumulative impact claim is prepared. The CVA and MVA are calculations of the contractor's overruns from its contract budget, detailed by cost code (typically the discipline of work performed, *i.e.*, civil, structural, mechanical equipment, piping, instrumentation, electrical, painting, insulation, etc.) and cost type (typically labor, materials, installed equipment, subcontract, field overhead/general conditions, etc.). The CVA and MVA are typically only slight modifications from a contractor's cost report. A summary-level CVA and MVA are shown in Table 5-1.

Table 5-1: Man-Hour/Cost Variances – Total Project

	Man-Hours	Labor Costs	Total Costs
Base Contract	120,000	$4,800,000	$10,000,000
Approved COs	30,000	$1,300,000	$2,000,000
Revised Contract	150,000	$6,100,000	$12,000,000
Actual MHs/Costs	200,000	$8,500,000	$15,000,000
Variance	50,000	$2,400,000	$3,000,000

The cost and man-hour data shown above, while hypothetical, is typical of highly impacted projects in that the contractor lost most of its money in labor. In this example, the contractor expended 50,000 more man-hours and $2.4 million more than its approved labor budget, which is likely due to a variety of factors including contractor bid error, contractor performance problems, and owner-caused problems. The contractor may have a basis for a cumulative impact of changes claim because it performed approved change order work which had an estimated value equal to 25

326 See *Wunderlich Contracting Co. v. U.S.*, 173 Ct.Cl. 180, 351 F.2d 956 (1965).

percent of its base contract man-hours, assuming that it did not include an estimate for loss of productivity for cumulative impact in its change order estimates, and that it reserved its rights to a cumulative impact claim.

Given a complex, highly impacted project, it is likely that other factors besides change orders affected the contractor's labor productivity. As one example, the contractor may have been forced to accelerate, subjecting the work to productivity losses as a result of extended overtime periods, overcrowding, inefficient crew sizes, and other factors.

A contractor seeking equitable compensation on its cost overrun has several damages methods from which to choose. Several variables factor into the decision, including:

1. The detail and reliability of the contractor's quantity, cost, and man-hour records. Were actual installed quantities, costs and man-hours accurately allocated and collected by day, week, or month in accordance with the project's work breakdown structure (WBS)? Were actual change order man-hours captured in separate cost accounts? Certain methods can be utilized only if this detail exists; however, this type of information is usually considered to be fundamental to the quantification and proof of compensable damages.

2. The detail captured in contemporaneous project documentation. Were detailed daily reports kept? Do they indicate the effects of owner-caused impacts? Does project correspondence adequately document owner-caused impacts? These records, demonstrating the cause of the problems and the effect on the contractor's productivity, are important to all methods. Still, to support detailed quantum calculations, contemporaneous documentation must be detailed as well.

3. The stage of the claim's process. Is the claim for negotiating purposes only, or will it be submitted as part of an expert report to the court? Certain methods are useful in negotiations, but are ineffective in court.

4. The size of the contractor's cost overrun. From a business perspective, the costs to prepare the damages analysis must be weighed against the size of the potential recovery.

5. The extent of contractor-caused problems and the ability to separate those costs. It is problematic in most methods if contractor-caused problems cannot be identified, segregated, and accurately priced.

Standard methods for calculating damages that include an element of productivity loss claims are discussed below in Section 5.2. In certain methods, all loss of productivity costs, including costs associated with the cumulative impact of changes and other impacting events, are subsumed into the total claimed costs. That is, costs associated with the basis of each issue, such as differing site conditions, delay, cumulative impacts, or acceleration, are not segregated. These methods include the Total Cost method, the Modified Total Cost method, the Jury Verdict method, and the *Quantum Meruit* method. Other productivity loss calculation methods, such as the Delta Estimate method, include a direct calculation of the costs being claimed for each claim issue, including claims for cumulative impact. The cumulative impact claim calculation methods employ MCAA factors, the Leonard Study models, the CII studies, and Ibbs models, or some combination of these studies. The Discrete Damages/Cost Variance Analysis method, which is described in Section 5.2.7, can include either a direct calculation of productivity losses using industry studies, the Measured Mile, Earned Value, or MCAA factors, or an indirect calculation where the contractor claims a portion of its cost overrun for productivity losses.

A contractor may also have been damaged not by the cumulative impact of changes, but rather by other specific issues which cause loss of productivity, such as directed acceleration or overtime. A brief overview of the studies that are used to show such impacts is also provided.

5.2 Methods of Damages Calculations

This section discusses and compares alternative methods for calculating and presenting damages, including: the Total Cost, Modified Total Cost, Jury Verdict, "A"/"B" estimates, *Quantum Meruit*, Delta Estimates, and the Discrete Damages/Cost Variance Analysis method. Also discussed are the criteria that are used to determine which one should be used.

5.2.1 Total Cost Method[327]

The Total Cost method involves a simple claim calculation based upon the assumption that all cost overruns are the result of the owner's actions. The contractor claims the difference between the costs it expended and the costs it was paid, and adds applicable overhead and profit. Referring to the project conditions in Table 5-1, the contractor could claim $3 million plus overhead and profit under the Total Cost method. Alternatively, the contractor could claim only its total labor losses of $2.4 million as a total labor claim. In either case, the claimed costs for cumulative impact of changes, and all other productivity losses, are subsumed into the total claim amount.

327 See Buoncrisitiani, David and Grove, Jesse B. "Defending Against The Contractor's Total Cost Claim," The Construction Superconference, December 1993.

The Total Cost method of presenting damages is merely a calculation of the difference between the contractor's total actual cost of performing the work, and the contractor's current contract value. The Total Cost method is illustrated by Figure 5-1.

Figure 5-1: Total Cost Method

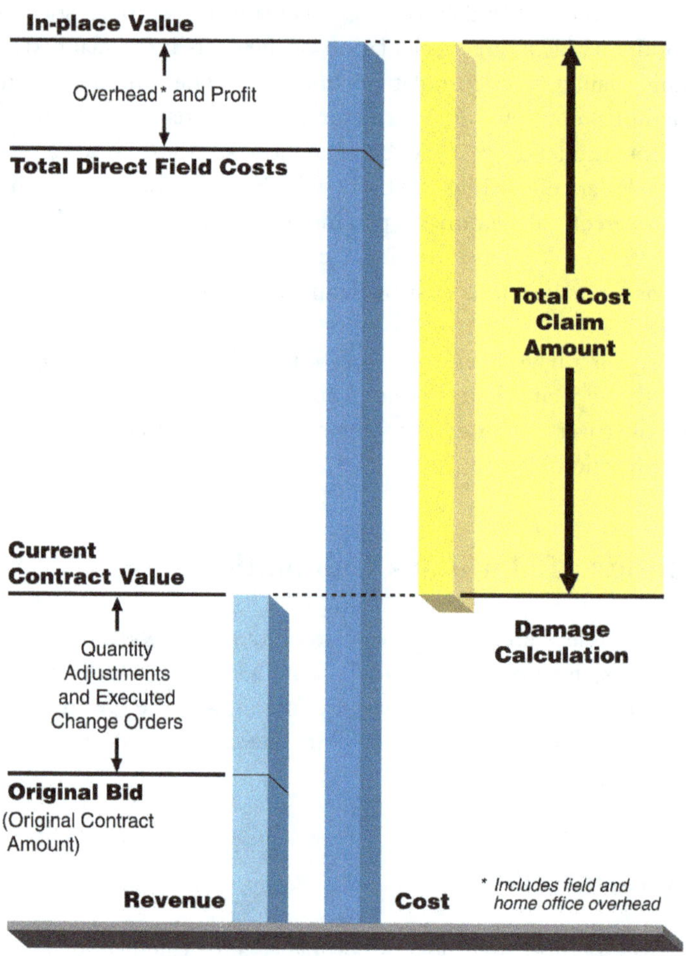

An early application of the Total Cost method of calculating contract damages in the United States refers to a 1951 decision by the then U.S. Court of Claims, which adopted a variant of Total Cost.[328] Over the ensuing years, this court occasionally permitted use of Total Cost, rejected its use on a number of occasions, and gradually clarified when it could properly be employed.

Other federal courts and state courts also have rendered decisions on claims involving the Total Cost calculation of damages, as discussed below. Courts have rejected the

[328] See *Great Lakes Dredge & Dock Co. v. United States,* 96 F. Supp. 923 (119 Ct.Cl. 1951).

use of the Total Cost method for factual, not legal, reasons. American courts rarely embrace Total Cost calculations, and do so only if certain elements of proof are set forth by the claimant. Courts outside the United States also disfavor Total Cost claims.

5.2.1.1 Elements of Proof

In 1968, the U.S. Court of Claims[329] set out the elements of proof that a claimant must demonstrate to prevail on a Total Cost claim for damages. The elements are:

> *(1) The nature of the [claimant's] particular losses make it impossible or highly impracticable to determine them with a reasonable degree of accuracy;*
> *(2) The claimant's bid or estimate was realistic;*
> *(3) Its [the claimant's] actual costs were reasonable; and*
> *(4) It [the claimant] was not responsible for the added expenses.*

The WRB court wrote:

> *This theory [total cost] has never been favored by the court and has been tolerated only when no other mode was available and when the reliability of the supporting evidence was fully substantiated.*[330]

Thus, Element No. 1 sets forth two requirements:

1. Total Cost is to be used only as a fail-safe; and
2. It must rest on reliable supporting evidence.

These requirements have been widely followed by other courts. The owner's classic argument is that the contractor has not shown that its excess costs were caused by the owner's breach, and that the contractor has only shown its final costs. If the contractor merely measures its cost overruns for the entire project by summing overruns from each work activity, the contractor is not providing proof that the costs were caused by the owner's breach; it is merely breaking the overruns into smaller components. Unless it can show some "Measured Mile" effect for each individual work activity, there is no proof that the cost overrun was caused by a particular breach. If a court determines that it is impractical to prove losses directly, then the court finds no "Measured Mile" method exists to ensure that the claim is reasonably certain. With respect to proof of causation, the casual connection that the overrun resulted from a specific breach is implied because of the difficulty of actual proof.

---

329 See *WRB Corp. v. United States*, 183 Ct.Cl. 409, 426-427 (1968) at paragraph 8.

330 *Id.*

Elements 2 through 4 above are derived from a 1955 Court of Claims' criticism of Total Cost method:

> *This method of proving damage is by no means satisfactory, because, among other things, it assumes plaintiff's costs were reasonable and that plaintiff was not responsible for any increases in cost, and because it assumes plaintiff's bid was accurately computed, which is not always the case, by any means.*[331]

Thus, in Element 2, courts are concerned that contractors may try to turn a loss into a gain when a contractor has underbid a job. The bid estimate is usually prepared by a contractor in support of its commercial tender for a project, and it forms the basis for the preparation of a control budget, which is used for project cost control. The bid estimate should include sufficient costs for project engineering, procurement, and construction. The bid estimate may also include start-up, pre-commissioning, and commissioning costs, if these activities are part of the contract scope of work. The estimate should be based upon the contractor's knowledge of, and experience on, similar construction projects, its proposed construction means and methods and work plan, its knowledge of the project scope of work and regulatory requirements, and its relationships and experience with subcontractors and suppliers.

The value and accuracy of the bid estimate is based upon, and should consider, uncertainties and assumptions related to project scope, duration, and labor productivity, and the status of various project development and definitional considerations known at the time the contractor tenders its commercial proposal. As such, the bid estimate should include sufficient allowances (design development allowances, for example) and contingencies to provide for estimate uncertainties.

For Element 3, a court provided a definition of the reasonableness of actual costs: *"[A] reasonable basis for computation and the best evidence which is obtainable under the circumstances of the case, and which will enable the trier to arrive at a fair approximate estimate of loss is sufficient proof."*[332] As stated earlier, damages from loss of productivity may be difficult to prove and, therefore, it may be difficult to establish the reasonableness of the actual costs. For example, in *Aetna Casualty and Surety Co. v. George Hyman Construction Co.*,[333] the subcontractor, represented in this case by the surety, was contracted to perform electrical work on an Amtrak rail station. The plaintiff claimed loss of productivity under the "cumulative impact" theory, but the court rejected the plaintiff's claim and stated that: *"[T]he mere expression of an estimate as to the amount of productivity loss by an expert witness with nothing to support it will not establish the fundamental fact*

331 *F.H. McGraw and Co. v. United States*, 130 F.Supp. 394, 397, 398, 400 131 Ct.Cl. 501 (1955).

332 *Bagwell Coatings, Inc. v. Middle S. Energy, Inc.*, 797 F.2d 1298, 1309 (5th Cir. 1986).

333 *Aetna Casualty & Surety Co. v. George Hyman Construction Co., U.S. Dist.*, LEXIS 22627 (E.D. Pa. 1998).

of resultant injury nor provide a sufficient basis for making a reasonable approximation of damages."[334]

Regarding Element No. 4, a 2002 California court held that the contractor failed to satisfy that it was not responsible for the increased costs.[335]

Thus, Total Cost may be used to ensure equity when the factual information is dependable and when no other method to calculate damages can be used. However, courts prefer the discrete calculation of damages:

> *The proper measure of damages is the amount of the contractor's extra costs directly attributable to the [owner's] breaches. Obviously, the preferable method for calculating such losses would be to itemize and total the cost of each piece of equipment or material and each man-hour necessitated by the unanticipated conditions encountered in performing the contract.*[336]

The Modified Total Cost method[337] has been developed to overcome these concerns. In a Modified Total Cost claim, the contractor first deducts its bid errors, its own or its subcontractor's performance problems and inefficiencies and rework costs, and damages that were not caused by the owner.[338]

5.2.1.2 Theoretical Bases

The Total Cost method is based on the premise that damages do not have to be calculated with exactitude, and on the courts' concern with ensuring that injured parties are compensated. A 1986 Pennsylvania court invoked a widely accepted American damage rule:

> *[M]ere uncertainty as to the <u>amount</u> of damages will not bar recovery where it is clear that damages were the <u>certain result</u> of [the owner's] conduct.*[339] [emphasis added]

334 *Id.* at 262 (citing *Luria Bros. & Co. v. U.S.*, 177 Ct. Cl. 676, 369 F.2d 701, 713 (1966)).

335 See *Amelco Elec. v. City of Thousand Oaks,* 98 Cal. Rptr. 2d 159 (Cal Ct. App. 2000).

336 *State Highway Comm'n of Wyoming v. Brasel & Sims Construction Co., Inc.,* 688 P.2d 871, 877 (Wyo. 1984). Also see the Delta Estimates Method and the Discrete Damages/Cost Variance Analysis method in Sections 5.2.6 and 5.2.7.

337 See Chapter 5, Section 5.2.2 herein.

338 See Sweet, *Legal Aspects of Architecture Engineering and the Construction Process,* p. 608 (4th ed.).

339 *Larry Armbruster & Sons, Inc. v. State Public School Bldg. Auth.,* 95 Pa. Commw. 310, 505 A.2d 313 (1986); *Pugh v. Holmes,* 486 Pa. 272, 405 A.2d 897 (1979).

Previously, a 1955 court wrote:

> *The plaintiff must prove its damages with reasonable certainty, but they do not have to be proved with absolute certainty. It is sufficient if plaintiff furnishes a reasonable basis for its computation, even though the result is only approximate.*[340]

To ensure that a claimant was not denied a recovery because of calculation problems, another court stated:

> *[Total cost method] is premised on the fact that where a contractor is entitled to an adjustment, the governmental body should not be exonerated merely because the contractor is unable to prove his increased costs with precision.*[341]

5.2.1.3 Prerequisites

Even when properly used, the Total Cost method is the last step, not the first or only step, in claim preparation. The Total Cost becomes a relevant issue only after the claimant has proved entitlement and cause-effect. Total Cost is a method for calculating damages, and not a method of proving liability.

> *The 'total cost plus profit' theory of computing damages...assumes that defendant is in fact liable for all the injuries sustained....*[342]

Most American courts disfavor the Total Cost method:

> *This measure of damages has not been favored by the courts.*[343]

> *Our opinion in Great Lakes Dredge & Dock Co. v. United States was not intended to give approval to this method of proving damage, except in an extreme case and under proper safeguards.*[344]

> *[W]e view basing damages on the difference between bid estimate and actual costs with trepidation.*[345]

---

340 *F.H. McGraw and Co. v. United States,* 130 F.Supp. 394, 397, 398, 400 131 Ct.Cl. 501 (1955). Also see Rubin, *The Total Cost Method of Computing an Equitable Adjustment – An Analysis,* 26 Fed.B.J. 303, 304 (1966).

341 *John F. Harkins Co. v. School Dist. of Philadelphia,* 313 Pa. Super. 425, 460 A.2d 260, 263 (1983).

342 *Wunderlich Contracting Co. v. U.S.,* 173 Ct.Cl. 180, 351 F.2d 956 (1965), at p. 12 (Westlaw print).

343 *John F. Harkins Co. v. School Dist. of Philadelphia,* supra, 460 A.2d at 263.

344 *F. H. McGraw & Co. v. United States,* supra, 130 F.Supp. at 400.

345 *Oliver-Finnie Co. v. United States,* 279 F.2d 498, 506, 150 Ct.Cl. 189 (1960).

5.2.1.4 Owner's Failure to Provide Alternate Method

Owners should consider the risk of an adverse decision when the contractor uses a Total Cost method to calculate its damages. In the alternative, an owner could present evidence that damages can be calculated on a discrete basis, and that Total Cost need not be used. If the owner takes the position, as a matter of law, that Total Cost should not properly be used, and does not offer an alternative discrete method of calculating damages, the court could rule that it had no other method to determine damages and accept the contractor's claim as calculated using the Total Cost method. Failure to provide an alternative damages method could make it easy for the courts, if entitlement and cause-effect evidence has been persuasive – and because they could be unsophisticated in construction cost accounting method – to decide that no other method but Total Cost was put forth in evidence to determine the damages. While it may be counterintuitive for owners to use an expert to calculate an alternate version of the contractor's damages, such testimony may be prudent to preclude the courts' acceptance of the contractor's use of Total Cost method.

An example of this concept is the 1987 case, *Pennzoil Co. v. Texaco, Inc.*[346] Although not a Total Cost construction claim, Pennzoil and Texaco had a dispute over the value of Getty Oil, which each wanted to purchase. Pennzoil sued Texaco for the tort of intentionally interfering with the Getty Oil-Pennzoil handshake deal prior to Texaco extending an offer to buy Getty Oil. Pennzoil had made its argument for damages as follows: Getty Oil owned a billion barrels of oil reserves. If Texaco had not interfered, Pennzoil would have purchased Getty Oil (and its oil reserves) for $3.4 billion. The average cost of finding new oil reserves in the United States at the time was $10.87 per barrel. Thus, to find a billion dollars' worth of new oil would have cost Pennzoil $10.87 billion. Pennzoil should, therefore, be entitled to the difference between the price it would have had to pay to find a billion barrels of oil, and the price it would have had to pay for Getty Oil.[347] However, the value of Getty Oil was very close to the price that Texaco had paid for it.[348]

Nevertheless, Texaco's attorney decided not to put on witnesses to rebut Pennzoil's damages evidence. The jury returned a verdict awarding Pennzoil $7.53 billion in compensatory damages, plus $3 billion in punitive damages. Had Texaco offered expert testimony as to an alternative value of damages, the jury may have considered it and awarded Pennzoil a lower amount.

---

346 *Pennzoil Co. v. Texaco, Inc.*, 481 U.S. 1, 4 (1987).

347 See Lloyd, Robert M., "Pennzoil V. Texaco, Twenty Years After: Lessons For Business Lawyers,"Transactions: The Tennessee Journal of Business Law, Vol. 6, 2005 at 346.

348 *Id.* at p. 347.

5.2.1.5 The Total Cost Cumulative Impact Claim

Like a Total Cost claim, the cumulative impact claim is often viewed with judicial circumspection because productivity can be affected by numerous contractor-caused or non-compensable events:

> *The "cumulative impact" claim is viewed with circumspection because productivity can be affected by numerous impacting events other than compensable events. The "cumulative impact" claim, essentially a "total labor cost" claim, requires proof that:*
>
> 1. *Impact attributable to changes was unforeseeable or was expressly excluded from change order settlements;*
> 2. *The changes were the sole cause of disruption for which the claim is made;*
> 3. *The "cumulative impact" was excessive and unreasonable in relation to what the contractor might have expected;*
> 4. *Impact costs cannot be segregated; and*
> 5. *Cumulative impact costs can be reasonably proven as to amount.*[349]

Therefore, a contractor can recover damages under the Total Cost method only if it can show that: (1) it was impracticable to prove actual losses directly; (2) its bid was reasonable; (3) its actual costs incurred were reasonable; and (4) it was not responsible for any of the additional costs that it incurred.

This finding requires that the contractor's pre-contract productivity estimate was accurate, the contractor did not cause the loss of productivity, and the contractor could not have foreseen and addressed cumulative impact in the pricing of its negotiated change orders.[350] Owners may argue successfully that changes, in cumulative effect, must fundamentally change the work so as to constitute a "cardinal change" to provide entitlement to a cumulative impact claim outside of the changes clause, as a breach of contract.[351]

Under the constructive change theory, the owner has the right to make changes under the terms of the contract, but has not followed the procedures of the "Changes" clause contained in the contract. Thus, the contractor argues that its contract was constructively changed. A cardinal change is a change that fundamentally alters the contemplated scope of work, and is a breach of contract which alters the work so drastically that the contractor is required to perform duties materially different from

---

349 See *Servidone*, 931 F.2d at 861-62. See also *Youngdale & Sons Const. Co., Inc. v. U.S.*, 27 Fed. Cl. 516 (1993) at p. 16 Westlaw (disallowing recovery under "total cost" method); also see *John E. Green Plumbing & Heating Co., Inc. v. Turner Const. Co.*, 742 F.2d 965, 968 (6th Cir. 1984) (affirming denial of unsegregated disruption claim).

350 Also see Chapter 2, Section 2.2.3 and Chapter 5, Section 5.2.1 herein.

351 See *L.K. Comstock & Co., Inc. v. Becon Const. Co., Inc.*, 932 F. Supp. 906 (E.D. Ky. 1993) for a more thorough discussion of the "cardinal change" doctrine.

those for which it contracted. The contractor generally bears a heavier burden of proof to recover under a cardinal change claim than when proceeding under a constructive change theory. The contractor must prove that the excessive change orders materially or fundamentally altered the scope of the work or nature of the bargain, such that the work no longer resembles the original contract.[352] For example, one court determined that the contractor did not prove that 65 change orders materially altered the nature of the bargain, and was therefore not entitled to costs that allegedly resulted from the effects of those changes.[353]

However, certain courts have found entitlement to a contractor's cumulative impact claim under the cardinal change doctrine.[354] In such cases, *"a cardinal change constitutes a breach of the contract and exposes the owner or construction manager to breach of contract damages, which can include anticipated profits and other remedies not traditionally provided for in contractual remedy granting provisions."*[355]

5.2.2 Modified Total Cost Method

The Modified Total Cost approach is a damages analysis method whereby the claim amount is produced by calculating the difference between an adjusted in-place value and an adjusted bid amount. The adjusted in-place value is determined by taking the project's in-place value (total field costs plus overhead, profit and bond) and adjusting it downward for identified contractor-caused performance errors and other noncompensable costs.[356] The adjusted bid amount is determined by adjusting the original bid (contract) amount for executed change orders and identified bid errors. As shown by Figure 5-2, the Modified Total Cost method is similar to the Jury Verdict method, except that the contractor presents the claim value by adjusting the total costs for noncompensable costs and bid errors, rather than the "jury" or "trier of fact" making these adjustments.

---

352 See Jones, Reginald M., "Lost Productivity: Claims for the Cumulative Impact of Multiple Change Orders," Public Contract Law Journal, Volume 31, Number 1, Fall 2001, p. 17, published by the Section of Public Contract Law, The American Bar Association.

353 See *Aragona Constr. Co., Inc v. United States,* 165 Ct. Cl. 382 (1964).

354 See, *e.g., Southwest Marine, Inc.,* DOTCAB No. 1663, BCA ¶27,102; *Dyson & Co.,* ASBCA No. 21,673, 78-2 BCA ¶13,482.

355 Sgarlata, Barnes & Brasco, "Successful Project Management through an Understanding of the Risks and Benefits of Early Claims Resolution," (CMAA 2004 National Conference), p. 7.

356 See *Great* Lakes *Dredge & Dock Co. v. United States,* 96 F. Supp. 923 (119 Ct.Cl. 1951); *Sovereign Construction Company, Ltd.,* ASBCA No. 17792, 75-1 BCA ¶ 11,251, at 53,606.

Figure 5-2: Modified Total Cost Method

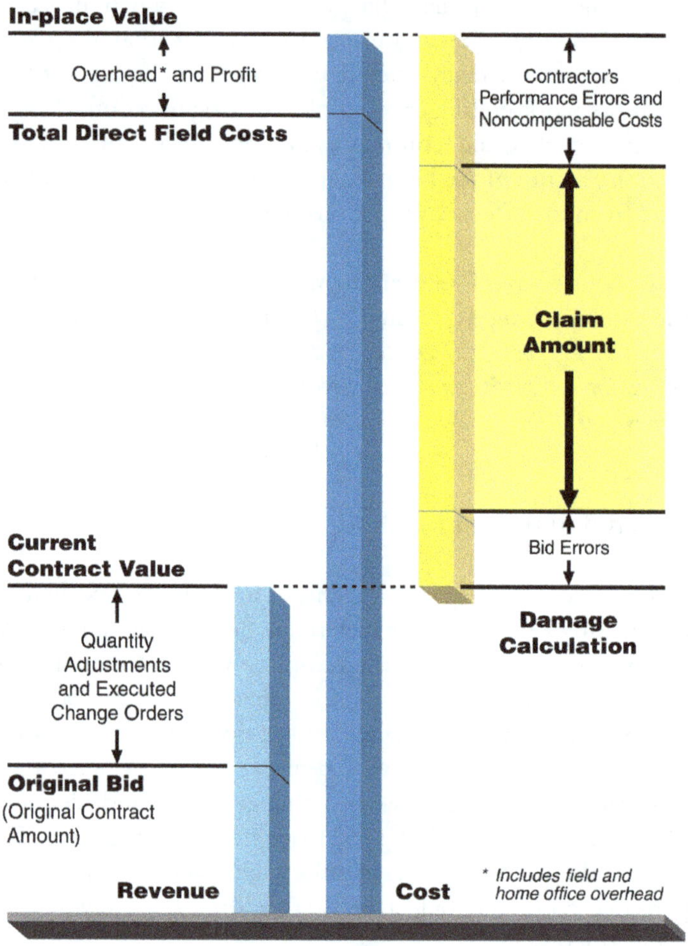

The Modified Total Cost method is held in higher regard than the Total Cost method, but the contractor must still satisfy the four elements applicable to the Total Cost approach,[357] namely the impracticability of proving actual losses with a reasonable degree of accuracy, the reasonableness of the bid/budget costs, the reasonableness of the contractor's actual costs, and the lack of contractor responsibility for additional costs.

---

357 See *AMEC Civil, LLC v. DMJM Harris, Inc.*. Civil Action No. 06-64 (FLW), US District Court, D. New Jersey (June 2009) (citing *Propellex,* 342 F.3d at 1339; *Servidone,* 931 F.2d at 861; and *S. Comfort Builders,* 67 Fed. Cl. 124, 139 (2005) 124, 139 (2005) at 146-47). See, also, *Youngdale & Sons Const. Co., Inc. v. U.S.*, 27 Fed. Cl. 516 (1993) at 541 (citing *Servidone,* 931 F.2d at 861): *"The modified total cost method is simply* the *total cost method* modifie*d or adjusted for any deficiencies in the plaintiff's proof in satisfying the four requirements of said method."*

Use of the Modified Total Cost approach is also advantageous to a contractor on those construction cases with schedule delays, lost productivity, and numerous unresolved claims that are mostly intertwined and contain inextricable elements of changes and impacts to the unchanged work. In these cases, it is usually impossible or highly impractical to keep accurate records by which the contractor can link the costs incurred to specific problems encountered on the job. In addition, this approach is best suited to those cases where it would be highly impractical to even attempt to compute damages by any other method. The contractor must, however, identify and accurately quantify its bid and performance errors to use this method.

Because the Modified Total Cost approach may achieve a higher claim amount than other methods, the owner may be disadvantaged. However, one advantage that the owner enjoys is that the contractor is usually required by the court to disclose its cost records. This affords the owner the opportunity to perform its own damage analysis in preparation for defense of the claim. By analyzing the contractor's actual cost records, the owner may be able to demonstrate additional cost overruns that are attributable to bid errors or contractor execution errors. The owner's analysis may also be able to show that certain cost accounts showing cost overruns have no relationship to owner-caused problems or, at least, both the owner and contractor impacted certain activities, for which the contractor is inappropriately blaming the owner for 100 percent of the cost overrun.

One board adopted a Modified Total Cost approach in *Robert McMullan & Sons, Inc.*, finding that the contractor's claim arose almost from the beginning of the contract, continuing and enlarging itself every day of the project. The board stated that to attempt to segregate delay and lost time, set up and break down estimates, and evaluate interferences and loss of efficiency as opposed to normal time would be a wasteful exercise, for in the end, the result should approach the total cost actually experienced. It was impractical, if not impossible, to distinguish with any degree of accuracy the contract work from extra work.[358]

This line of reasoning was also evidenced by another board in its decision regarding the appeal of *Sovereign Construction Company, Ltd.*:

> *Inefficient work is an intangible commodity. It is not possible to make a direct measurement of it. In identifying and measuring inefficiency a comparison must be made to some accepted standard. In this appeal we are unaware of any better method by which the appellant could establish additional costs attributable to inefficiency than by establishing what the work reasonably should have cost and what the work did in fact cost. Of course, the actual costs are subject to reduction if they include elements for which the Government is not responsible.*[359]

---

358 See *Robert McMullan & Sons, Inc.*, ASBCA No. 19129, 76-2 BCA ¶ 57,947.

359 *Sovereign Construction Company, Ltd.*, ASBCA No. 17792, 75-1 BCA ¶ 11,251 at 53,606.

One criticism sometimes proffered by an owner in a complex loss of productivity case may be that the contractor ought to have performed the changed or additional work in a different manner, and should have managed the problem less expensively by performing in a different fashion.

This Monday-morning quarterbacking has not received the support of the decisional law. In *Great Lakes Dredge & Dock Co. v. United States,*[360] the government urged that the contractor ought to have used a different method of solving a difficult problem for which the government had payment responsibility under the contract. The court said that it was not interested in a hypothetical alternative solution.[361] The court then adopted a Modified Total Cost approach, stating:

> *In order to show what its increased costs were the plaintiff has produced proof to show what its estimated costs were and what its actual costs were. The difference it has adjusted by certain errors that it admits it had made in making up its estimate, and also certain costs which it admits were not attributable to the encountering of this subterranean water.*[362]

The court found that the contractor had underbid the project and, accordingly, the court adjusted the plaintiff's bid. The court then compared the adjusted bid to the cost of performing the work. The court made adjustments for costs incurred that were not the government's fault, as suggested by the contractor.[363]

In another instance the Court of Claims was faced with a case in which the plaintiff contractor relied on a Modified Total Cost approach. The court stated:

> *In essence, what plaintiff seeks to recover as an equitable adjustment is the difference between the total costs it actually incurred and what it 'should have' cost to produce the 200,000 MK-8 cartridges had there been no defective specifications.*
>
> *Teledyne is confronted with an accusation that some of its costs should be disallowed. However, no reason whatsoever has been given as to why these costs should be disallowed. Other than the defective specifications, all other possible reasons for these additional costs have been eliminated. In requiring plaintiff to prove still more (more than it has already proven), in the absence of evidence by the Government as to why the costs should be disallowed, the Board, in effect, applied a standard of proof approaching "beyond a reasonable doubt."*[364]

---

360 *Great Lakes Dredge & Dock Co. v. United States,* 96 F. Supp. 923 (119 Ct.Cl. 1951).

361 *Id.*

362 *Id.*

363 *Id.*

364 *Teledyne McCormick-Selph v. U.S.*, 588 F.2d 808, 810 (Ct.Cl. 1987).

The court accepted the approach and held for the plaintiff contractor.

The Modified Total Cost approach has met the approval of the boards and courts. Although it is not mathematically certain, such certainty is not required. The general rule regarding whether exact ascertainment of damages is required was articulated by the Court of Appeals for the Ninth Circuit in the case of *Elete, Inc. v. S.S. Mullen, Inc.:*

> *The difficulty of ascertainment of the amount of damage is not to be confused with the right of recovery. The rule is that, if the plaintiff produced the best evidence available and if it is sufficient to afford a reasonable basis for estimating his loss, he is not to be denied a substantial recovery because the amount of the damage is incapable of exact ascertainment.*[365]

This principle was also discussed by a New York court in *S. Leo Harmonay Inc. v. Binks Manufacturing Company*:

> *It is fundamental to the law of damages that one complaining of an injury has the burden of proving the extent of harm suffered; delay damages are no exception.... On the other hand, courts have often recognized that the extent of harm suffered as a result of delay, such as the loss of efficiency claim in issue, may be difficult to prove. Thus, courts have recognized that a plaintiff may recover even where it is apparent that the quantum of damage is unavoidably uncertain, beset by complexity, or difficult to ascertain, if the damage is caused by the wrong. The law is realistic enough to bend to necessity in such cases.*[366]

Damages need not be proven with mathematical certainty.[367] The Modified Total Cost approach utilizes the best available information to determine as closely as possible what the damages should be. *"Moreover, evidence of damages may consist of probabilities and inferences."*[368] The ASBCA has noted:

> *Where, however, the contractor has proved that it suffered extra expenses by reason of changes, delays by the Government, etc. the mere fact that the amount of an equitable adjustment cannot be computed with mathematical certainty does not necessarily defeat recovery. If a fair and reasonable approximation can be determined, the Courts and this Board have done so.*[369]

A 2009 New Jersey court categorically rejected a general contractor's attempt to use the Modified Total Cost method to prove its delay damages in claims asserted against

---

365 *Elete, Inc. v. S.S. Mullen, Inc.*, 469 F.2d 1 127 (1972) at 1131.

366 *S. Leo Harmonay Inc. v. Binks Manufacturing Company*, 597 F. Supp. 1014 (S. D. N.Y. 1984) aff'd. 762 F.2d. 990 (2d Cir. 1984).

367 See *E.C. Ernst, Inc. v. Koppers Co., Inc.*, 626 F.2d 324, 327 (3rd Cir. 1980).

368 *E.C. Ernst, Inc. v. Koppers Co., Inc.*, 626 F.2d 324, 327 (3rd Cir. 1980).

369 *Drexel Dynamic Corp.*, 67-2 BCA ¶ 6,410, pages 29,698, 29,699 (ASBCA 1967).

the design consultant on a public construction project. The court similarly rejected the contractor's attempt to use a cumulative impact theory of causation, which the plaintiff described as a "ripple effect" attributable to numerous design deficiencies, finding that there were "substantial questions" regarding the contractor's ability to show causation. While the court concluded that the contractor could proceed to trial on other aspects of its claim, the court cautioned that it would not allow the contractor to avoid its own negligence, or other potential causes of delay and damage unrelated to the designer's negligence, by asserting that there was a "cumulative impact" of delay and damages that was solely the designer's fault.[370]

However, a 2010 California Court of Appeal case unequivocally establishes the Modified Total Cost theory of damages as a viable remedy for California contractors.[371] The appellate court found that where no contractual requirement exists for a contractor to document its actual costs, that contractor may be able to recover for work it performed in good faith using the Modified Total Cost method. During the subject wastewater treatment plant project, the owner issued over 300 change orders containing more than 1,000 changes to the plans and specifications. The appellate court found that the Modified Total Cost theory was valid, depending on interpretation of the contractual requirements for documentation of the actual costs of change orders. Under the Total Cost method, damages are determined by subtracting the contract amount from the total cost of the contractor's performance. Under the Modified Total Cost method, if the contractor is responsible for some of its increased costs of performance, then those costs are subtracted from the contractor's damages to arrive at the Modified Total Cost value. The court found that the contractor was not contractually required to document its actual costs (which the contractor contended was not possible), and that the contractor could use engineering estimates to prove its claims under a Modified Total Cost method, provided those estimates were the best evidence available.

In 2011, the Washington Court of Appeals upheld the Modified Total Cost method of calculating damages.[372] The court determined that the contractor's computation of damages using the Modified Total Cost approach was appropriate, because the contractor's expert considered the cost for which the contractor was responsible and the financial impact of the delay that the contractor had to the project. The contractor had also shown that its bid was reasonable.

---

370 See *AMEC Civil, LLC v. DMJM Harris, Inc.*. Civil Action No. 06-64 (FLW), US District Court, D. New Jersey (June 2009).

371 See *Dillingham-Ray Wilson v. City of Los Angeles* (2010) 182 Cal. App.4th 1396.

372 See *Coastal Constr. Grp., Inc. v. Stellar J. Corp.*, Court of Appeals of Washington, Division One, No. 66932-0-1, October 2011.

5.2.3 Jury Verdict Method[373]

The Jury Verdict method provides the courts another way to calculate damages when discrete/actual damages are not presented by the contractor or owner. The term "Jury Verdict" is misleading in that most cases that have received damages awards on a Jury Verdict basis involve decisions by courts or Boards of Contract Appeals, not juries. In these cases, judges and board panel members serve the roles of jury and triers of fact, hearing testimony and weighing evidence, and awarding damages based on an approximation of damages suffered.

Needles v. United States in 1944 was one of the earliest Jury Verdict cases.[374] The Jury Verdict method has been applied in many subsequent cases. Examples of courts and boards justifying the Jury Verdict method follow:

> *[I]t would be unjust to only permit recovery for a recognized wrong by holding plaintiff to a near impossible standard of requiring him to list with mathematical precision the amount of every item of damage flowing from that wrong.*[375]

> *When a contractor has proven entitlement, but cannot define his costs with exact data..., it is not uncommon to resort to the 'jury verdict' method of determining excess costs. It is not essential that the amount be ascertainable with absolute exactness or mathematical precision. It is enough if the testimony and evidence adduced is sufficient to enable the court or board (acting as the jury) to make a fair and reasonable approximation of the amount recoverable.*[376]

> *When a contractor has proven entitlement, but cannot define his costs with exact data, courts and appeal boards have been reluctant... to send the contractor away empty handed when it is at all possible to make a fair and reasonable approximation of the extra costs.*[377]

> *When a contractor has proven entitlement, but cannot define his costs with exact data, courts and appeal boards have been reluctant... to send the contractor away empty handed when it is at all possible to make a fair and reasonable approximation of the extra costs.*[378]

> *Where a court or a board of contract appeals sitting as the finder of fact is confronted with competent but conflicting evidence on what is a reasonable amount for an equitable adjustment, it may employ what is referred to as the 'jury verdict approach.'... The*

---

373 See Buoncrisitiani, David and Grove, Jesse B., "Defending Against The Contractor's Total Cost Claim," The Construction Superconference, December 1993.

374 See *Needles v. United States,* 101 Ct. Cl. 535 (1944).

375 *Fattore Co., Inc. v. Metropolitan Sewerage Comm'n,* 505 F.2d 1, 4 (7th Cir. 1974).

376 *E. Arthur Higgins,* 79-2 BCA ¶ 14,050, pp. 69,052, 69,066 (AGBCA 1979) .

377 *Id.*

378 *Id.*

conceptual touchstone for the use of the jury verdict approach is the existence of conflicting competent evidence calling into question the accuracy and reliability of the plaintiff's computations. As the name implies, the technique by necessity requires the application of the fair, equitable and informed discretion of the trial judge.[379]

Courts prefer that damages be proven discretely:

The proper measure of damages is the amount of the contractor's extra costs directly attributable to the government's breaches. Obviously, the preferable method for calculating such losses would be to itemize and total the cost of each piece of equipment and each man hour necessitated by the unanticipated conditions encountered in performing the contract.[380]

The U.S. Court of Appeals for the Federal Circuit explained this preference:

Clearly, the actual cost method is preferred because it provides the court, or contracting officer, with documented underlying expenses, ensuring that the final amount of the equitable adjustment will be just that – equitable – and not a windfall to either the government or the contractor.[381]

However, courts and boards have found that unless the Jury Verdict approach, with its relaxed evidentiary requirements, is used, the contractor often would be unable to recover.[382] If proof of actual costs associated with the breach of contract events that were experienced by the contractor is not presented, courts may choose the Jury Verdict method instead of accepting the Total Cost method. A 1985 Arizona case expressly prohibited the use of Total Cost *"when there is sufficient evidence to use the jury verdict method."*[383] However, another court found that the Jury Verdict method *"is not favored and may be used only when other, more exact, methods cannot be applied."*[384] In 2000, the board in *Clark Construction Group, Inc.*[385] awarded cumulative impact damages to the contractor and its subcontractors using a Jury Verdict method, based on MCAA productivity loss factors as a measure of productivity loss.

Courts are concerned with the "Jury Verdict" method because, as one court put it: *"its primary peril, as evidenced in this case, is the risk that unrealistic assumptions will be adopted and*

---

379 *Delco Electronics Corp. v. United States*, 17 Ct.Cl. 302, 320-324 (1989) at 323.

380 *State Highway Comm'n of Wyoming v. Brasel & Sims Construction Co., Inc.*, 688 P.2d 871, 877 (Wyo. 1984).

381 *Dawco Constr., Inc. v. United States*, 930 F.2d 872, 880, 882 (1991).

382 See *R-D Mounts, Inc.*, 75-1 BCA ¶ 11,237, Pages 53,491, 53,492 (ASBCA 1975).

383 See *New Pueblo Constructors, Inc. v. State,* 144 Ariz. 95, 696 P.2d 185, 194 (1985). See also *Metropolitan Sewerage Comm'n v. R.W. Constr.*, 78 Wis.2d 451, 255 N.W.2d 293-302 (1977).

384 *Dawco Constr., Inc. v. United States*, 930 F.2d 872, 880, 882 (1991).

385 See *Clark Construction Group, Inc.*, 00-1 BCA ¶ 30,870, VABCA No. 5674 (April 2000).

extrapolated, greatly multiplying an award beyond reason, and rewarding preparers of imprecise claims based on undocumented costs with unjustified windfalls."[386]

5.2.3.1 When Can the Jury Verdict Method be Used?

The Jury Verdict method may be used when the following three well-established requirements are met:

(1) That clear proof of injury exists;
(2) That there is no more reliable method for computing damages; and
(3) That the evidence is sufficient for a court to make a fair and reasonable approximation of the damages.[387]

5.2.3.1.1 Proof of Injury

A contractor must establish both the fact and the amount of injury. This requires a contractor to demonstrate that the owner is liable for the damages that it incurred.[388]

The contractor must then quantify its damages. After entitlement for damages is factually established, courts *"are not over-strict in demanding specific proof of exact amounts."*[389] Damages need not be calculated with *"absolute exactness or mathematical precision."*[390] *"It is sufficient if he furnishes the court with a reasonable basis for computation, even though the result is only approximate."*[391] However, the contractor has the burden *"to show a sound basis for the quantum of his claims by a preponderance of credible evidence."*[392] Failure to prove the entirety of the damages claimed does not bar recovery of lesser amounts when properly established.[393]

---

386 *Mergentime Corp. v. WMATA,* 1997 U.S. Dist. LEXIS 23408 at 5.

387 *Dawco Constr., Inc. v. United States,* 930 F.2d 872, 880, 882 (1991) at 880; See also *Delco Electronics Corp. v. United States,* 17 Ct.Cl. 302, 320-324 (1989); *and WRB Corp. v. United States,* 183 Ct.Cl. 409, 426-427 (1968).

388 *See Delco Electronics Corp. v. United States,* 17 Ct.Cl. 302, 320-324 (1989). Thus, the contractor must prove liability, causation and resultant injury; *Wunderlich Contracting Co. v. U.S.,* 173 Ct.Cl. 180, 351 F.2d 956 (1965).

389 *Bell v. United States,* 404 F.2d 975, 983, 186 Ct.Cl. 189, 205 (1968).

390 *Specialty Assembling & Packing Co. v. United States,* 355 F.2d 554, 572, 174 Ct.Cl. 153 (1966).

391 *Wunderlich Contracting Co. v. U.S.,* 173 Ct.Cl. 180, 351 F.2d 956 (1965), at p. 16 (Westlaw print).

392 *Greenwood Constr. Co., Inc.,* 78-1 BCA ¶12,893, pp. 62,818, 62,830-1 (AGBCA 1977).

393 See *Bell v. United States,* 404 F.2d 975, 983, 186 Ct.Cl. 189, 205 (1968).

5.2.3.1.2 No Other Reliable Method of Calculating Damages

As a second requirement, the contractor must then demonstrate that another, more precise method of calculating damages is not available:

> *[T]he contractor bears the burden of establishing that no more reliable method is available than the 'guesstimate' of the 'jury verdict method', i.e., a method that would more precisely calculate the cost of the extra work.*[394]

An Arizona court held that the Jury Verdict approach cannot be used *"where there is no excuse for the failure to keep track of actual costs."*[395] However, the court upheld a Jury Verdict award where *"due to circumstances beyond his control, the claimant cannot feasibly prove specific damages by a more reliable method."* When liability to contractors is clear, owners should consider presenting their own damage calculations. Otherwise, the court may start with the contractor's calculation of damages, which may be flawed.

5.2.3.1.3 Adequate Evidence

Finally, the contractor must ensure that the evidence is adequate. The record must contain a basis for a fair and reasonable approximation of damages before the court can adopt a Jury Verdict award.[396] However, courts may, in part, base a Jury Verdict award on estimates prepared by experts, comparative cost information from similar projects, or reductions from total cost accounting information.

5.2.3.2 Procedure

The Jury Verdict approach generally involves two steps. In the first step, the court determines that the owner breached the contract and the contractor is entitled to an equitable adjustment.[397] In the second step, the court determines the amount of compensable damages, deriving a total value for the claim or element of a claim, often expressed in a lump sum. In computing this value, the court typically deducts, from

394 *Dawco Constr., Inc. v. United States*, 930 F.2d 872, 880, 882 (1991).

395 *New Pueblo Constructors, Inc. v. State,* 144 Ariz. 95, 696 P.2d 185, 194 (1985).

396 See *California Canners & Growers Ass'n v. United States,* 9 Ct.Cl. 774, 785 (1986).

397 The vast majority of cases employing the jury verdict method involve equitable adjustment claims. Federal government contracts and some state government contracts contain equitable adjustment clauses. These clauses provide administrative remedies to compensate contractors for changes that otherwise would amount to breaches of contract requiring litigation to resolve. See, *e.g., Fattore Co., Inc. v. Metropolitan Sewerage Comm'n,* 505 F.2d 1, 4 (7th Cir. 1974). As a BCA explained: *"[A]n equitable adjustment is a corrective measure which is simply intended to keep the contractor whole when the contract has been modified by the Government."* *Greenwood Constr. Co., Inc.*, 78-1 BCA ¶12,893, pp. 62,818, 62,830-1 (AGBCA 1977). Thus, equitable adjustments generally serve the same purpose as ordinary contract damages.

the contractor's overall cost overrun or request for damages, an amount that it believes was caused by the contractor's performance problems, and potentially another amount attributable to bid error. The remainder is the Jury Verdict award of compensable damages. Figure 5-3 illustrates the concept of the Jury Verdict approach to calculate damages.

Figure 5-3: Jury Verdict Method

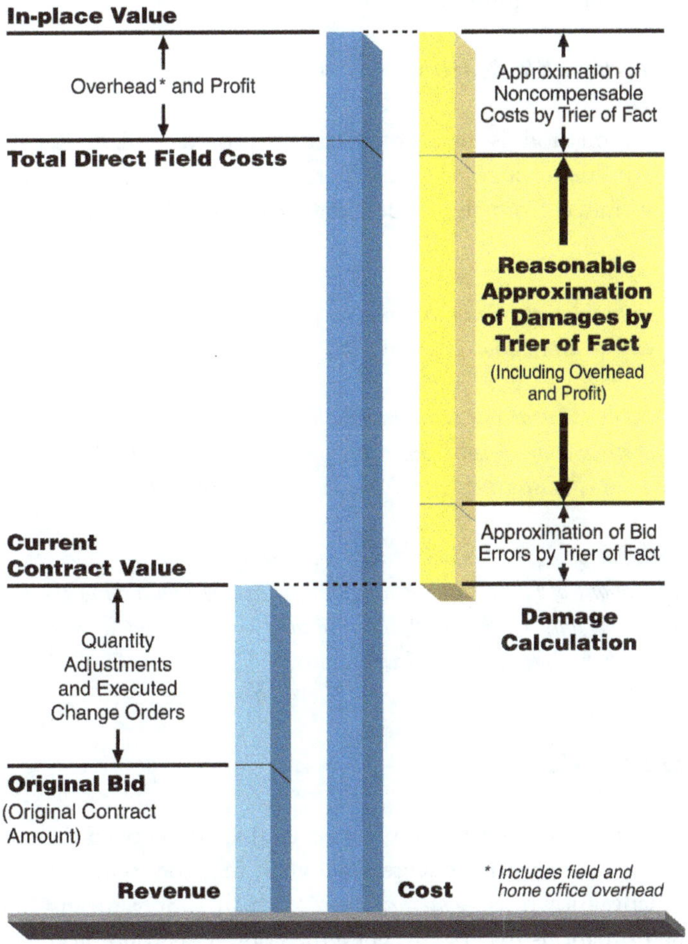

5.2.3.3 Reasonableness Requirement

Damages must be reasonable. The nature of the changes requested by the owner may, among other things, influence a court's assessment of reasonableness when using the Jury Verdict method. If changes required performance well beyond what was originally contemplated under the contract, or required a higher degree of skill and expertise, courts may expand the range of reasonable recovery. A Court of Claims wrote:

> *It has been recognized that a contractor's costs will be considered reasonable in spite of production difficulties and delays where a government change order demands a much higher degree of technical skill and expertise than would have normally been expected under the original contract.*[398]

In contrast, when a contractor agrees to manufacture an advanced product and assumes the risk of difficulties in design and production, large cost overruns may not meet the reasonableness requirement.

5.2.3.4 Unpredictable Nature

The Jury Verdict method is neither precise nor predictable. Legal, factual, and equitable factors influence a court's Jury Verdict analysis. The court's "gut reaction" or "an inference drawn from the evidence as a whole"[399] may guide the amount of the award:

> *In using the jury verdict approach, the processes leading to the amount of adjustment are not readily susceptible to articulation in an opinion.*[400]

> *Our judgment is based upon our familiarity with the records in the two cases, our knowledge of legal precedents and the general experience we have gained in deciding from day to day matters arising out of construction as well as other types of contracts.*[401]

> *The determination of a reasonable recovery, it seems, can only be made by reference to contemporary notions of fairness and equity, and by taking into account the totality of the circumstances. The duty of the court then, is to determine what a reasonably prudent contractor would have done under the circumstances.*[402]

5.2.4 *Quantum Meruit*[403]

For various reasons, the contractor may be unable to recover damages under the original contract. When this occurs, recovery may be pursued under a quasi-contractual remedy known as *Quantum Meruit*, which simply means "as much as he deserves." Such situations may occur when the contract is found to be unenforceable, or when the contractor performs extra work that is not covered by the original

398 *Delco Electronics Corp. v. United States*, 17 Ct.Cl. 302, 320-324 (1989).

399 *Western Contracting Corp. v. United States*, 144 Ct.Cl. 318, 320, 333, 334, 369 (1958).

400 *Drexel Dynamic Corp.*, 67-2 BCA ¶ 6,410, pages 29,698, 29,699 (ASBCA 1967).

401 *Phillips Construction Co. v. United States*, 394 F.2d 834, 839, 184 Ct.Cl. 249 (1968).

402 *Delco Electronics Corp. v. United States*, 17 Ct.Cl. 302, 320-324 (1989).

403 See Buoncrisitiani, David and Grove, Jesse B. "Defending Against The Contractor's Total Cost Claim," The Construction Superconference, December 1993.

contract. Where *Quantum Meruit* recovery is sought, the contractor will set forth its determination of the reasonable value of the benefit conferred upon the owner. All competent evidence of actual expenditures of the contractor will usually be considered by the court in determining that reasonable value.

A contractor may recover under a theory of *Quantum Meruit* if it can prove substantial performance on the construction contract, and an endeavor on its part in good faith to perform fully under the contract. An intentional departure from the requirements of the contract is *per se* not in good faith, unless the failure is: (i) caused at least in part by the defendant, or (ii) the departure is so trifling so as to be *de minimis*. Additionally, a party seeking recovery under a theory of *Quantum Meruit* which has been terminated wrongly is excused from proving substantial performance of the entire contract, but it must prove substantial performance and an intention to completely perform up until the date of termination.

Quantum Meruit may apply when an express contract is abandoned. Courts will seek a remedy such that the contractor is to be paid the reasonable value of its work. The rules governing when an abandonment of contract occurs to enable recovery in *Quantum Meruit* include:

> *Rescission by abandonment requires mutual assent of the parties.*[404]
>
> *An abandonment may be accomplished by express mutual consent or by implied consent through the actions of the parties.*[405]
>
> *A contract will be treated as abandoned when the acts of one party, inconsistent with its existence, are acquiesced in by another....Where acts and conduct are relied upon to constitute abandonment, however, they must be positive, unequivocal, and inconsistent with intent to be further bound by the contract.*[406]
>
> *Uncommunicated subjective intent by one party to abandon is not sufficient to release the obligations.*[407]
>
> *If there is no special agreement, the builder's compensation is the reasonable value of the services rendered and the materials furnished by him; and this is so, also, where the*

---

404 *City of Del Rio v. Ulen Contracting Corp.*, 94 F.2d 701, 704 (5th Cir. 1938). Accord, California Civil Code § 1689 (a).

405 *Schwartz v. Shelby Constr. Co.*, 338 S.W.2d 781, 788 (Mo. 1960).

406 *H.T.C. Corp. v. Olds*, 486 P.2d 463, 466 (Colo.App. 1971).

407 *Modern Builders, Inc. of Tacoma v. Manke*, 27 Wash.App. 86, 615 P.2d 92 (1980).

building plan is abandoned to such an extent that it is impossible to trace the contract in the work done.[408]

In *C. Norman Peterson Co. v. Container Corp.*,[409] *Quantum Meruit* was used to calculate the reasonable value of the work performed by the contractor resulting from numerous change orders and contract abandonment. The California state court found that the scope of work had been changed due to an excessive number of changes, and invoked the cardinal change doctrine to allow recovery on a *Quantum Meruit* basis.

There are three prerequisites to establishing abandonment of contract and obtaining recovery in *Quantum Meruit*:

1. There was "substantial" change in the work;
2. The change was not covered by the contract or contemplated by the parties; and
3. The contractor should not have anticipated or discovered the change.

Typically, under *Quantum Meruit* recovery, value is calculated in terms of the contractor's cost of performance rather than by some other measure, such as the increase in value of the owner's property.[410] Even when the contract is not abandoned, a contractor can recover in *Quantum Meruit* for extra work outside the scope of the contract.

5.2.5 "A"/"B" Estimates

The "A"/"B" Estimate method is generally utilized in cases where the additional work and the cause of the delay or disruption is fairly straightforward. For example, a lump sum contractor bids on a chemical plant expansion, assuming that a large pressure vessel is to be located at grade with its own foundation. After the job begins, the contractor discovers that the engineer has relocated the vessel in the structure above grade because of other process changes made by the owner. The contractor in such a case can prove his damages by subtracting an independent estimate of the work as planned (the "A Estimate") from the new estimate of the work with the vessel relocated (the "B Estimate"). The "B Estimate" is a fair and reasonable estimate of the value of the work, at the time of the bid, if the contract documents had contained an adequate description of the problems and changes that actually occurred during the course of the work.

---

408 *Obert v. Ede,* 38 Wis.2d 240, 156 N.W.2d 422, 424 (1968), quoting 17A Corpus Juris Secundum, Contracts § 367(1), p. 374.

409 See *C. Norman Peterson Co. v. Container Corp of Am.*, 218 Cal Rptr. 592 (Cal. Ct. App 1985).

410 See *City of Portland ex rel. Donohue & Fleskes Corp. v. Hoffman Constr. Co.,* 236 Ore. 739, 596 P.2d 1305, 1309, 1313 Fn. 22 (1979).

The contractor's damages under this method would be the difference between the two estimates, as shown by Figure 5-4. By following this simple approach, the contractor is able to eliminate from the damages any costs not directly related to the vessel relocation itself, such as labor inefficiency not caused by the change. The approach is advantageous because of its simplicity, but is difficult to use where several different factors affecting the contractor's performance are part of the job. Its conclusions may be deemed speculative, and care must be taken to avoid absurd damage results. Care must also be taken to ensure that the estimates are reasonable and consistent in method.

Figure 5-4: "A"/"B" Estimate Method

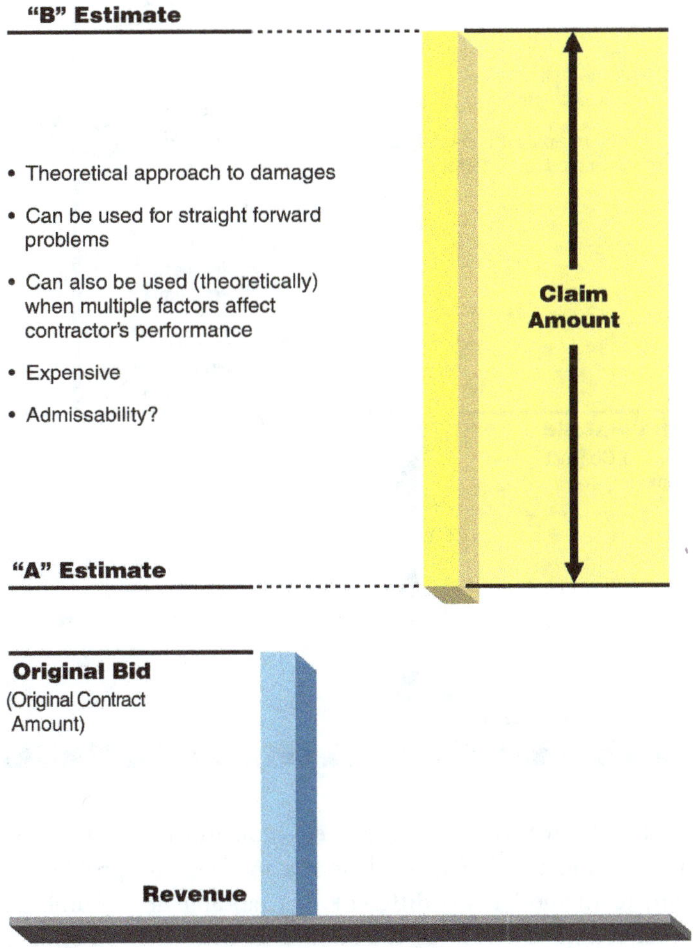

5.2.6 Delta Estimate Method

For each of several distinct problems that occur on a project, specific cause-effect relationships and estimates of damages or actual costs are established using the Delta Estimate method, as shown by Figure 5-5.

Figure 5-5: Delta Estimates Method

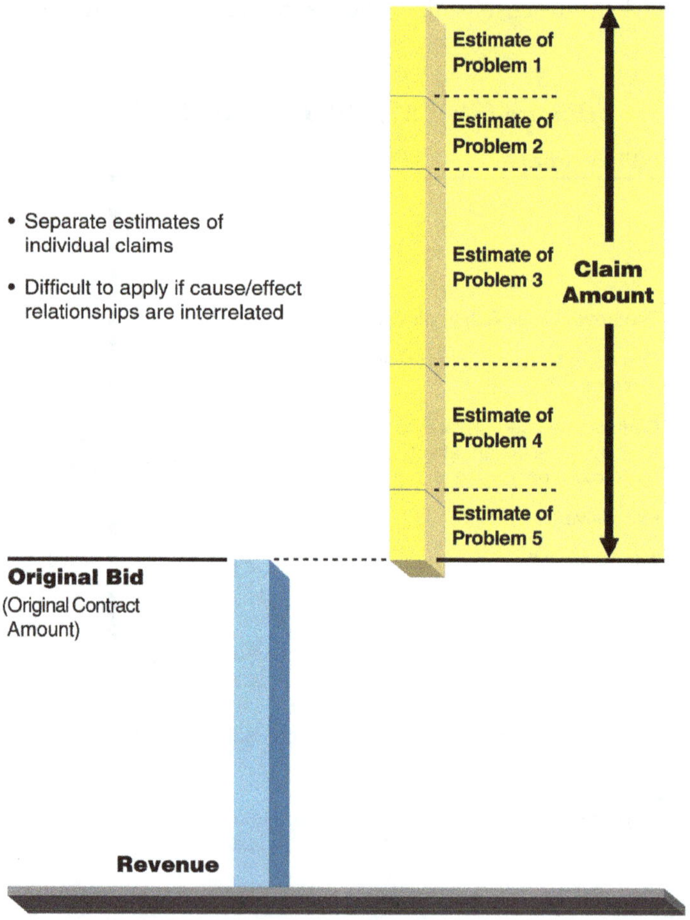

For this method to be feasible, cost documentation must be kept by the contractor that clearly distinguishes the damages relating to the separate problems. Simply said, the Delta Estimate method is no different than adding up a number of individual change order requests. The Delta Estimate becomes difficult to apply if several problems and their effects are interrelated. Also, the Delta Estimate method is often criticized because the sum of the parts may exceed the whole (cost overrun). For example, contractors often calculate overall increases in labor costs by using multiple causes of lost productivity, such as:

- Excessive and sustained overtime;
- Stacking of trades;
- Overcrowding;
- Effect of multiple changes;
- Temperature; and
- Others.

Each of these causes of productivity loss is evaluated, and the calculated loss from each cause is summed to yield the overall productivity loss claim. However, without comparing the resulting total labor increase to the actual labor man-hour variance between the contractor's control budget and the total actual labor man-hours expended, including consideration for approved and pending changes, the calculation of the sum of the parts may yield a result that is greater than the whole variance. An owner should insist on a review of the contractor's job cost report to evaluate the actual variance between the control budget man-hours and actual labor man-hours for individual work activities. These problems are overcome by using the Discrete Damages/Cost Variance Analysis method of calculating damages (see Section 5.2.7).

5.2.7 Discrete Damages/Cost Variance Analysis Method

By applying a variation of the Delta Estimate and Modified Total Cost methods, the claimant can go one step further to provide convincing proof of the damages incurred using an approach known as the Discrete Damages/Cost Variance Analysis method, which is illustrated by Figure 5-6.

The method involves the specific distribution of all costs incurred on the project, rather than quantifying only the claimed costs, as may be done in the other methods. The credibility of this method is established by separating the cost growth that results from bid error, non-compensable problems, and compensable problems, and then identifying individual compensable problems with specific costs tied to each problem. In addition, the cost growth for each discrete claim is applied to each relevant cost account (or group of accounts) to demonstrate that the "sum of the parts" of each claim does not exceed the entire cost overrun for each cost account.

The analyst implementing this method, in addition to evaluating cost growth for each cost account (or group of accounts), may consider quantity, man-hour, unit price, and unit productivity variances for each account based upon a comparison of the current budget and actual values for these incises, to the extent that such data is available. Because of the comprehensive nature of this type of analysis, it relies heavily upon the contractor's contemporaneous detailed man-hour and cost records and reports, and the current budget versus actual quantity, man-hour, and cost data derived for each account.

Figure 5-6: Discrete Damages/Cost Variance Analysis Method

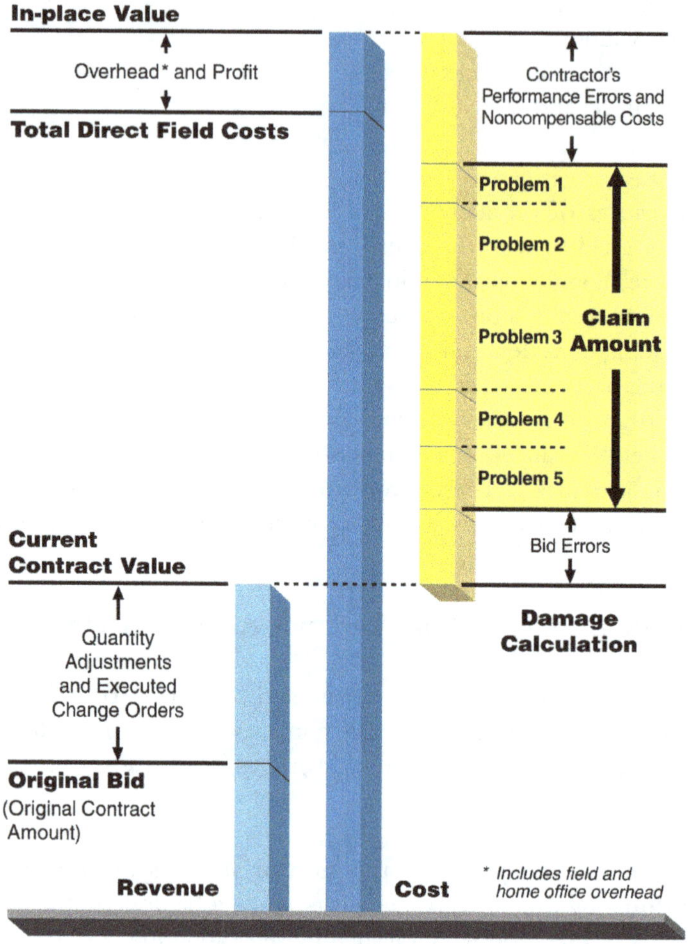

In the Discrete Damages/Cost Variance Analysis method, one appropriates all costs taken from the claimant's cost records into cost variance analysis categories, which set forth the claimed costs separate from the other non-compensable causes of cost overruns. These cost variance analysis categories include the original bid, any bid errors, approved changes, pending changes, specific compensable problems (claims), and non-compensable (or unrequested) costs. Because the claimant is in each case associating a particular item of damage for each compensable problem, the cause-effect relationship is established as the numbers are calculated. Accordingly, the claimed amount is both easy to verify and difficult to refute. Most contractors maintain cost records that have some degree of subdivided cost account detail. Therefore, it is more credible to demonstrate variances at the greatest level of detail possible using the available cost records.

The analyst may employ varying types of analyses and cost apportionments when there are several interrelated problems, such as a project impacted by both numerous owner-caused changes and contractor-caused bid errors and performance problems. For example, assume that a contractor has experienced a cost overrun of $100,000 across four separate cost accounts associated with piping labor. In one analysis, the contractor quantifies a bid error of $10,000 related to missing scope of work. In a second analysis, the contractor determines that $15,000 of piping labor is contained within disputed change order requests. In a third analysis, the contractor uses industry studies to quantify $50,000 of direct disruption associated with overtime and overcrowding. [411] In a fourth analysis, the contractor quantifies $5,000 of increased piping labor costs due to overtime labor rate premiums. Of the $100,000 lost, the contractor has accounted for $80,000, of which $70,000 will be claimed. If the extent of project change was significant enough to support a cumulative impact claim, some or all of the residual $20,000 in these accounts could be claimed. If not, the contractor could still demonstrate that it is not claiming more than it lost.

This method may be difficult to apply in a case where there are several interrelated problems, but any attempt to segregate and price claims separately will generally increase the chance of recovery. Because of the detailed analyses involved, this method can be more time-consuming and costly to prepare than other methods, and it can only be implemented if the contractor has maintained detailed cost and man-hour records.

The advantages of the Discrete Damages/Cost Variance Analysis method are:

1. Because all components of the cost records are evaluated and distributed into the claim matrix, the total value of the requested funds can be fully explained and justified;
2. The link between entitlement and the amount of damages is developed at the lowest level of detail of the cost records, thus providing significant credibility to the burden of proof; and
3. Isolation of the damages of cost components and individual claim problems identifies the hard areas of claim, while also identifying the soft spots to be negotiated, if necessary.

5.2.7.1 Nine Step Process

The Discrete Damages/Cost Variance Analysis method involves nine steps.

Firstly, by analysis of the contractor's detailed cost records, each project cost account or work activity is divided into basic cost elements such as quantities; man-hours; direct labor and burden costs; construction equipment ownership and operating costs;

---

411 See Sections 3.4.2 herein for a discussion of these industry studies.

permanent materials and equipment costs; subcontracts and supplier costs; rental equipment and operating costs; consumable materials and supply costs; field overhead costs; and home office overhead costs. Profit may not be shown in the contractor's job cost reports, and may need to be determined from its internal bid estimate or other cost accounting reports. However, profit needs to be addressed in the overall cost accounting, such that the contractor's bid estimate costs plus anticipated profit comprises the original total contract price.

A typical cost account structure is shown by Figure 5-7. If the project is still ongoing, actual progress information through the date of the actual costs is used to develop earned budget values, so that the budget estimate and actual cost bases are similar. The bid cost (budget) and the actual cost for each cost element of each project work activity are identified.[412]

Figure 5-7: Typical Cost Account Structure

Work Activity No. and Description				Direct Labor		Construction Equipment		Perm Mat'l & Equipment		Subc.	Rental Equipment		Consumables	Field Overhead		Home Office Overhead		
No.	Description	Quantity	Unit	Hrs	$	Hrs	$	Hrs	$	$	Hrs	$	$	Hrs	$	Hrs	$	Total

Secondly, all approved changes to the contract price are applied to derive a current contract value for each work activity. Profit included in each approved change is allocated to the profit account. Noncontroversial pending changes that will comprise additional compensation to the contractor are also identified, and the cost accounts associated with these pending changes are populated with agreed costs for these pending changes, including overhead and profit.

The third step involves evaluation of the contractor's bid estimate to determine whether it underestimated any cost elements of any work activity. This evaluation comprises at least a review of quantities, unit productivity rates, and unit prices for accuracy and reasonableness, and adjustments for bid errors are quantified accordingly. These bid error costs are allocated to the contractor. Knowledge of industry standards, efficient construction techniques, man-hour productivities, and cost estimating are essential to properly evaluate the contractor's bid.

---

412 Often, the cost accounts used by the contractor during its bid preparation may be different than those that it used during the execution of the project in its job cost report. Hopefully, the contractor has reconciled the different cost accounts between the bid phase and the execution phase when it set up its job cost report. If not, the analyst will need to perform this reconciliation to prepare the cost/damages matrix.

As a fourth step, each activity and its quantity, labor man-hours, and cost elements are evaluated for non-compensable cost items. These may include costs incurred as a result of the contractor's performance errors and rework, labor strikes for which the contractor is responsible or has the contractual risk, idle time due to bad weather for which the contractor has the contractual risk,[413] equipment down-time costs caused by the contractor's improper operation or maintenance, inefficient management of the contractor's labor or the labor incurred by subcontractors, pending changes for which the contractor has no entitlement, or other cost items that are clearly not the owner's responsibility. Certain of these non-compensable costs may need to be estimated if the contractor's cost records are not sufficiently detailed to identify these costs.

As a fifth step, each work activity is assessed for the direct impact and cost overruns caused by each specific compensable claim problem. Damages such as additional labor, equipment, or material costs are identified and assigned to the appropriate claim problem.

As a sixth step, the detailed project as-planned and as-built schedules are analyzed to identify the time frame of specific delays, disruptions, changes, labor strikes, suspensions, adverse weather conditions, late deliveries of critical equipment and material, acceleration, and other delay problems. The time frame for each problem isolates when each problem and its associated cost occurred, which can then be traced back to the contractor's job cost reports and other man-hour records. Each cost account is evaluated for delay and acceleration impacts using appropriate schedule analyses that quantify delay and acceleration time and responsibility. Each cost account is assessed by time frame for direct impacts and cost overruns caused by discrete delay issues. Delay damages for extended time-related costs, standby costs, financing, and labor and material escalation costs for accounts such as field overhead staff, temporary services, and construction equipment are quantified and allocated to each party, based upon contribution to delay in each schedule period. Specific home office time-related costs are also allocated based on each party's responsibility for delays. Acceleration costs for additional equipment and supervision; direct labor cost growth for premium time and shift differentials; and man-hour growth due to inefficiencies resulting from common acceleration direct disruption impacts such as extended overtime, manpower density, and multi-shift coordination are quantified. Acceleration cost growth is allocated to the contractor and owner based upon contribution and responsibility for delay, and the subsequent acceleration. Cost overruns outside the time frame in which the problem or its effect occurred are excluded from the compensable cost distribution, and therefore are identified as non-compensable costs.

In the seventh step, all residual costs not accounted for in steps two through six are examined to determine whether they could have resulted from the indirect

[413] Unusually severe weather costs may be a compensable problem.

disruption/cumulative impact caused by changes and the other issues thus far evaluated and quantified. Residual costs that could not have resulted from indirect disruption/cumulative impact or one of the previously evaluated compensable issues and causes, such as unaccounted-for permanent material costs, are assigned to the contractor. Residual costs, and in particular residual man-hours, that could have resulted from indirect disruption/cumulative impact, and one or more of the previously evaluated issues, are quantified and distributed to the owner or contractor based on an assessment of impact using a Measured Mile analysis, industry study information, or contribution based upon direct scope and disruption man-hours quantified in the previous steps.

In the eighth step of the analysis, the cost/damage analysis matrix of the damage analysis components, project work activities, and cost accounts are totaled to determine the specific dollar amount of each requested claim (compensable problem). The actual cost for each project work activity by row is first distributed into one or more of the appropriate cost accounts. The actual cost in each cost account of each activity is then distributed into the appropriate damage analysis component category.

In this presentation, the actual cost variance or overrun in each cost account has been distributed into the appropriate damage analysis issue category, and all cost account variances have been distributed. The Cost/Damage Analysis Matrix is illustrated by Figure 5-8.

As a final, ninth step, the claimant must clearly communicate the basis of the distribution of costs into the claim categories, and support each element of the requested compensable damages with a cause-effect explanation to support the entitlement for cost recovery. See Chapter 7 for a discussion of how to link a cause-effect analysis to the allocation of damages.

5.2.7.2 Development of Cost Analysis Databases

The structure of the method first requires the establishment of two separate databases: a control budget database, and an actual cost database, which are linked by work activity and cost element definition. A third database, called cost variance/allocation, is developed from the first two.

The control budget is a restatement of the original bid estimate in a form that facilitates comparisons between budget and actual costs. The control budget is normally prepared after bid opening, and before construction starts.

Project Work Activities	Original Contract Value	Approved Change Orders	Current Contract Value	Pending Change Orders	Bid Error (Noncomp)	Contractor-Caused Problems (Noncomp)	Potential Compensable Damages: Issue No. 1	Issue No. 2	Issue No. 3	Issue No. 4	Total Actual Costs
ACTIVITY 100 – EARTHWORK	**$200,000**	**0**	**$200,000**	**0**	**0**	**$20,000**	**0**	**0**	**0**	**0**	**$220,000**
Direct Labor	40,000	0	40,000	0	0	0	0	0	0	0	40,000
Labor Burdens	15,000	0	15,000	0	0	0	0	0	0	0	15,000
Permanent Material & Equip	0	0	0	0	0	0	0	0	0	0	0
Consumable Materials	15,000	0	15,000	0	0	0	0	0	0	0	15,000
Owned Construction Equipment	30,000	0	30,000	0	0	0	0	0	0	0	30,000
Rental Equipment	20,000	0	20,000	0	0	0	0	0	0	0	20,000
Subcontracts	80,000	0	80,000	0	0	20,000	0	0	0	0	100,000
ACTIVITY 200 – CONCRETE	**$50,000**	**$20,000**	**$70,000**	**0**	**0**	**0**	**$30,000**	**0**	**0**	**0**	**$100,000**
Direct Labor	10,000	0	10,000	0	0	0	0	0	0	0	10,000
Labor Burdens	5,000	0	5,000	0	0	0	0	0	0	0	5,000
Permanent Material & Equip	0	0	0	0	0	0	0	0	0	0	0
Consumable Materials	5,000	0	5,000	0	0	0	0	0	0	0	5,000
Owned Construction Equipment	5,000	0	5,000	0	0	0	0	0	0	0	5,000
Rental Equipment	5,000	0	5,000	0	0	0	0	0	0	0	5,000
Subcontracts	20,000	20,000	40,000	0	0	0	0	0	0	0	40,000
ACTIVITY 300 – STRUCTURAL STEEL	**$400,000**	**$100,000**	**$500,000**	**0**	**0**	**0**	**$400,000**	**0**	**0**	**0**	**$900,000**
ACTIVITY 400 – INSTALLED EQUIP	**$4,000,000**	**$50,000**	**$4,050,000**	**0**	**0**	**$50,000**	**0**	**0**	**0**	**0**	**$4,100,000**
ACTIVITY 500 – PIPING	**$2,000,000**	**$100,000**	**$2,100,000**	**$30,000**	**$100,000**	**0**	**0**	**$300,000**	**$100,000**	**$200,000**	**$2,830,000**
...											
ACTIVITY 2100 – FIELD INDIRECTS	**$1,500,000**	**$50,000**	**$1,550,000**	**$5,625**	**$18,750**	**$13,125**	**$80,625**	**$56,250**	**$18,750**	**$37,500**	**$1,780,625**
ACTIVITY 2200 – HOME OFFICE OH	**$300,000**	**$10,000**	**$310,000**	**$1,125**	**$3,750**	**$2,625**	**$16,000**	**$11,250**	**$3,750**	**$7,500**	**$356,000**
ACTIVITY 2300 – FEES / PROFIT	**$200,000**	**$16,500**	**$216,500**	**$1,850**	**$6,125**	**$0**	**$26,000**	**$18,375**	**$6,125**	**$12,250**	**$287,225**
TOTAL PROJECT	**$8,650,000**	**$346,500**	**$8,996,500**	**$38,600**	**$128,625**	**$85,750**	**$552,625**	**$385,875**	**$128,625**	**$257,250**	**$10,573,850**

POTENTIAL CLAIM VALUE = $1,324,375

TOTAL COST OVERRUN = $1,577,350

Figure 5-8: Cost/Damages Analysis Matrix

5.2.7.2.1 Control Budget Calculations

The control budget is not the contractor's bid. The bid proposal is the price sheet used in the contract as the basis of payment for work performed. The contractor prepares an estimate of costs, overheads, and anticipated profit (the total of which becomes the bottom line of its bid), but the makeup of individual cost elements becomes hidden when assigning values (prices) to each bid item. Bid unbalancing is common. In addition, payment is usually based on units of work completed, and not for interim steps along the way. An example of the latter is payment for concrete - paid for by the cubic yard, in place. No separate payment is usually made for forming, pouring, finishing, etc. For this reason, the control budget (which shows separate estimated costs of forming, finishing, and placement of concrete), and not the bid schedule, is to be used as the basis for comparison for calculations of damages using the Discrete Damages/Cost Variance Analysis approach.

Thus, the control budget is a restatement of the original bid estimate and commercial proposal, in a form that facilitates detailed cost account comparisons and reporting between budget and actual costs, quantities, man-hours, and unit productivity rates. The control budget is normally prepared after the contractor's commercial tender is accepted by the owner, and before project-specific engineering and construction begins on a turnkey engineering, procurement, and construction (EPC) project. The control budget delineates the contractor's total estimated costs (by cost elements such as labor and material), quantities, and man-hours into cost accounts comprising the project scope of work defined by the project work breakdown structure (WBS).

A properly constructed control budget database is fundamental to the preparation of project cost and productivity reports to help control project costs and man-hours. The control budget is also used as the basis of comparison to actual costs and man-hours for variance analyses and damage calculations.

During claim preparation, particularly for disruption claims, control estimate quantities, man-hours, and costs are compared to actual information, so that planned and actual productivity rates for commodities and disciplines can be evaluated to determine the extent, nature, and responsibility for man-hour growth attributable to loss of productivity and disruption by cost account. In addition, the control estimate is evaluated for bid errors, which should be quantified before productivity loss calculations are performed. In the context of labor man-hour growth, the evaluation of the control estimate for bid errors usually includes a review of quantities, man-hours, unit productivity rates, allowances and contingencies, means and methods, estimate assumptions, and the project schedule.

The comparison of planned and actual labor productivity, and the evaluation of the control estimate for labor-related bid errors, is fundamental to the preparation of valid loss of productivity claims based on disruption or cumulative impact.

Control budgets usually vary over time, as illustrated by Figure 5-9.

Figure 5-9: Control Budget Revisions with Time

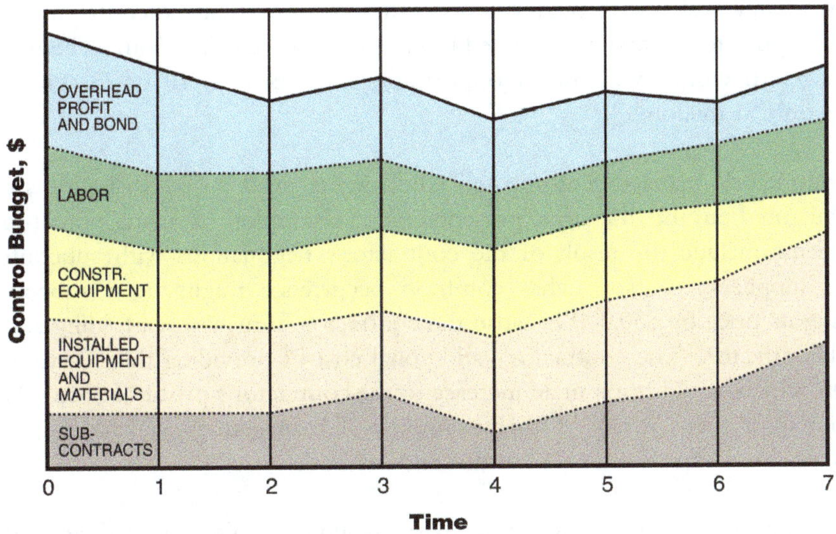

Various events cause this variance to happen; the most common being pay quantity variations, change orders, contractor-initiated improvements to means and methods, and subcontractor and supplier negotiations, as shown by Figure 5-10.

Figure 5-10: Sample Evolution of Control Budget

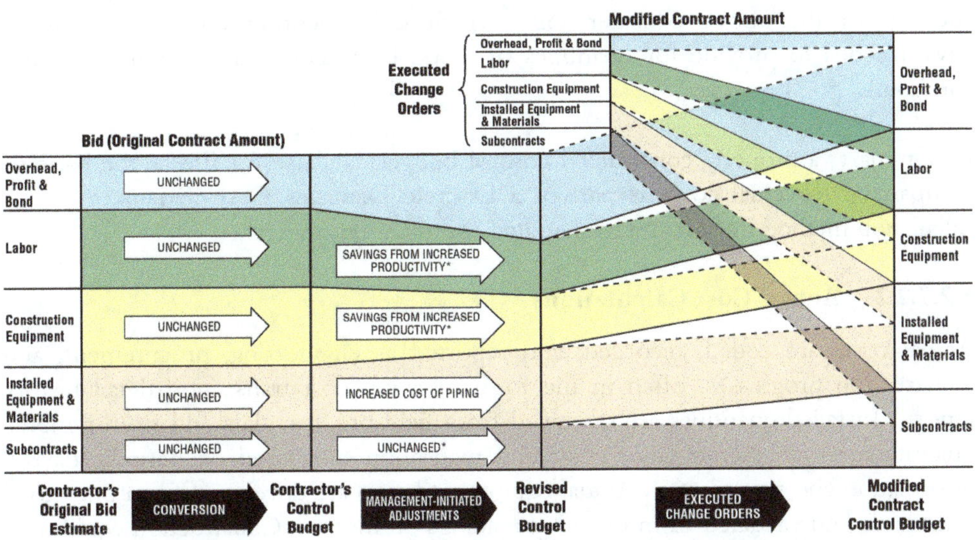

A restatement of the original bid estimate usually performed prior to construction to a form that facilitates comparisions between budget and actual costs.

* An example might be the result of negotiations with subcontractors and/or suppliers. A second example (graphically depicted above) occurs when management finds a better method of construction which is intended to save costs and result in increased profit such as purchase of a scaffolding system, the additional cost of which is more than offset by improved labor and equipment productivity.

In constructing a control budget database, the time dimensions must be included. In one case, a contractor submitted a Modified Total Cost claim of $4 million. The owner's outside auditor validated the loss at $4 million. When time-related cost reports became available through discovery, it was found that the contractor had lost $3 million prior to the first event stated to have caused the delay claim. The result was that the contractor's "window of opportunity" for recovery of its losses suddenly became only $1 million.

The contractor's management-initiated adjustments need to be well-addressed and well-structured in the damages presentation. Examples of management-initiated adjustments include the result of the contractor's negotiations with subcontractors and/or suppliers. A somewhat common occurrence might be a subcontractor dropping its price by $50,000 because the contractor furnished scaffolding service to the subcontractor. The contractor's additional cost of providing this service may be $15,000, which would result in an increase to the contractor's profit of $35,000 (which the contractor then owns). Other examples of management-initiated adjustments include post-bid improvements in means and methods.

The structure of the control budget database, including original, revised, and modified contract elements, must be time-variance oriented and trackable for categories of events. Typically, a contractor estimates the cost of the project on a two-dimensional spreadsheet, showing work activities on one axis and cost elements on the other. To provide for the complete comparison to actual costs, a third dimension – time – needs to be considered. Each event that changes the control budget is expressed as an additional spreadsheet containing the cost revisions to the preview control budget. In addition to the above considerations, which only mention costs, it is usually advantageous to provide for quantities of work, labor hours, and equipment types and hours.

In summary, a properly constructed control budget database for use as the basis of comparison to actual costs as part of a Discrete Damages/Cost Variance Analysis calculation method must be fully integrated and time sensitive.[414]

5.2.7.2.2 Actual Cost Calculations

Actual costs are coded, recorded, and reported as engineering, procurement, and construction progresses, often in the form of a job cost transaction detail history report. Installed quantities are compiled from field progress data and from periodic quantity surveys. Actual labor costs and man-hours are based on periodic payroll information and time sheets. Actual equipment hours are developed from operators' time cards and validated from recorded readings on meters. Construction equipment

414 The concept of a properly constructed control budget database is also applicable to the other damages calculation methods.

charges are based on hours of operation. Actual material and subcontract costs are captured and coded from invoices. The key to organizing actual cost and other data for comparison to the control budget is the proper and timely coding and posting of actual quantities, man-hours, and costs to the same account structure and level of detail that is defined by the control budget. In this way, total budget, earned budget, and actual costs, quantities, and man-hours can be compared and reported at the cost account level of detail.

Correlation and integration of cost accounts to an appropriate level of activity definition in the project schedule also facilitates project control, and can assist in the identification and quantification of cause-effect cost-schedule relationships for project issues such as delay and acceleration. If this correlation is performed well, subsequent cost and damage analysis work becomes a far simpler task. If it is not done well, the effort is large and the resulting cost becomes one that is not normally recoverable in a claim.

For damage analysis and quantification purposes, a properly constructed actual cost database should also be fully integrated and time sensitive. Actual costs are generally posted to the project cost system when they are processed for payment. This means that the actual costs from certain sources, such as material, outside equipment rental, and subcontractor invoices, may have been incurred one to two months earlier than the posting date. Therefore, if the costs are to be used in a time-sensitive analysis, one must ensure that proper adjustments are made to the actual cost database prior to comparisons which form the basis of damage calculations, to reflect when the work was actually performed.

Sometimes cost accounting adjustments are required; for example, when costs are miscoded and posted to an incorrect period of time. A case of miscoding is illustrated by a contractor's claim of $200,000 for owner-caused productivity impact on installing large bore pipe. The contractor's original bid estimate and control budget for the work was $500,000. The actual cost, as posted, was $700,000. The contractor's bid estimate and control budget for placing the pipe hangers was $100,000. The actual cost report showed no cost for the pipe hangers. Review of the project history through document analysis and interviews revealed that the foreman responsible for the pipe installation was also responsible for the pipe hangers. All of the pipe hangers had, in fact, been placed. The foreman had not separated the pipe hanger's effort, but had recorded all efforts to the piping installation task. The result was the contractor's claim was reduced to $100,000 ($200,000 piping installation overrun less $100,000 adjustment for the pipe hangers).

In addition to the above considerations, it is usually desirable to make a cost comparison on both actual quantities of work and Earned Value. In summary, a properly constructed actual cost database should be fully integrated and time sensitive.

5.2.7.2.3 Cost Variance Calculations

Cost variance calculations are simply the mathematical differences between budget elements and actual costs, preferably computed for increments of time and in total. By considering all project costs in the variance analysis, the analyst is afforded the opportunity to identify anomalies, such as the pipe hangers issue discussed earlier, and make appropriate cost accounting adjustments. Proper adjustments make the cost variance analysis more credible.

The cost report is the primary report used for project cost management and control, and in the context of the identification and analysis of budget estimate indices and cost overruns or variances, for claim calculations. This report compares information contained in the control estimate to information collected in the actual cost transaction system, and the report is prepared periodically (usually monthly) as progress is achieved and incurred costs are collected. This report usually presents total budget, earned budget, actual information, and sometimes forecast data, for each WBS cost account to which budget and/or actual costs have been coded. The cost report usually identifies earned budget versus actual quantity, man-hour, and cost element variances. The report is used contemporaneously to identify variances that require attention and corrective actions, and is often used for variance analyses and cost allocations for claim damages calculations

The use of a cost variance analysis is a fundamental component of the contractor's damages calculations on a complex engineering, procurement, and construction project, where the contractor is attempting to recover its increased costs for numerous problems affecting multiple disciplines of work. A contractor's failure to utilize this approach could severely limit the credibility of its claim because:

- The absence of support for the contractor's actual costs at the detailed cost account level in its cost reporting system may question whether the contractor actually incurred cost overruns in the areas of work that it alleged it was damaged;

- The absence of its bid estimate details, including the assumption it made in preparing its bid, may question whether the contractor is attempting to recover for its bid error;

- By not addressing the cost variance at the detailed cost account level, and not demonstrating that the sum of all of cost increases for each cost account, including those cost increases for which it seeks additional compensation, the contractor fails to demonstrate that its claim does not equate to more than the costs that the contractor incurred for each cost account; and

- The contractor has not addressed costs that resulted from its performance problems in addition to costs for which it is seeking additional compensation.

A cost variance analysis must often be supplemented by a quantity[415] variance analysis to ensure that:

- Increases in man-hours and related costs are not the result of quantity growth due to the contractor's bid error;

- The contractor can separate quantities that are included in approved and pending change orders, and not comingle the man-hours required to install those quantities with man-hours due to productivity loss;

- The unit man-hours to design or install different types of engineering or construction quantities can be separately applied during the variance analysis calculations;

- The contractor has considered quantity overages due to over-procurement and waste;

- Material and equipment costs for each quantity type can be applied during the variance analysis calculations to ensure accuracy of the results; and

- The sum of all of the quantities for each of the damages analysis categories does not exceed the total quantities installed by the contractor for each discipline of work.

A cost variance analysis must often be supplemented by a man-hour variance analysis to ensure that:

- Man-hours associated with quantity growth are not included in the contractor's other claims, such as claims for productivity loss;

- Man-hours included in pending and approved change orders are separately addressed and not included in the contractor's other claims, such as its change order and productivity loss claims;

---

415 Quantities include but are not limited to engineering deliverables, such as the number of design drawings, piping isometrics, and other documentation required by engineering to support the construction work, as well as physical quantities of materials to be installed during construction, such as concrete, soil quantities for import or export during site preparation, number and/or length of piles for foundations, structural steel, installed equipment, piping, instruments, electrical materials, etc.

- Man-hour growth that resulted from the contractor's bid error are not improperly included in the contractor's other claims;
- Man-hours resulting from contractor-caused problems are not included in the contractor's other claims; and
- The sum of all of the man-hours for each of the damages analysis categories does not exceed the total man-hours incurred by the contractor for each discipline of work.

The MCAA also recommends that the impact of changed conditions be quantified by the use of variance analyses:

> *For example, analyzing the amount of planned versus actual labor overtime on a project and then isolating when the overtime was occurred and what caused the overtime to be incurred can provide very useful information when evaluating the impact of changed conditions.*[416]

The use of a man-hour variance analysis can help ensure that a cumulative impact claim is not being used to capture increased man-hours that may have resulted from other causes. While the quantification of increased man-hours can be performed using various methods, such as a Measured Mile analysis or from using various industry studies, the man-hour variance analysis will frame the man-hours such that over-recovery is precluded, because the sum of the parts cannot exceed all of the man-hours incurred. If the man-hour variance is performed for all engineering and construction disciplines, significant credibility is built into the contractor's claim. If the contractor fails to use a man-hour variance analysis in its claim calculations, and the owner can demonstrate that the contractor is claiming more man-hours than it incurred when all other damage analysis categories are considered, the contractor's credibility suffers and its claim may be denied.

5.2.7.2.4 Allocation of Cost Variances

The final step in calculating damages under the Discrete Damages/Cost Variance Analysis method is to allocate the variances to the responsible party. Categories for allocation usually include contractor bid error, contractor performance error, other non-compensable costs, and contractor compensable claim costs, as shown by Figure 5-11. Identification and utilization of other categories of cost variances is commonly performed.

Allocation is made by expert opinion utilizing the cost variance analyses and other parallel analyses that may include the following:

---

416 "Change Orders Productivity Overtime: A Primer for the Construction Industry," MCAA, 2012, p. 7.

- Analysis of bid requirements;
- Analysis of contract requirements;
- Analysis of construction requirements;
- Schedule/delay/impact analysis; and
- Analysis of project history.

Figure 5-11: Allocation of Cost Variances

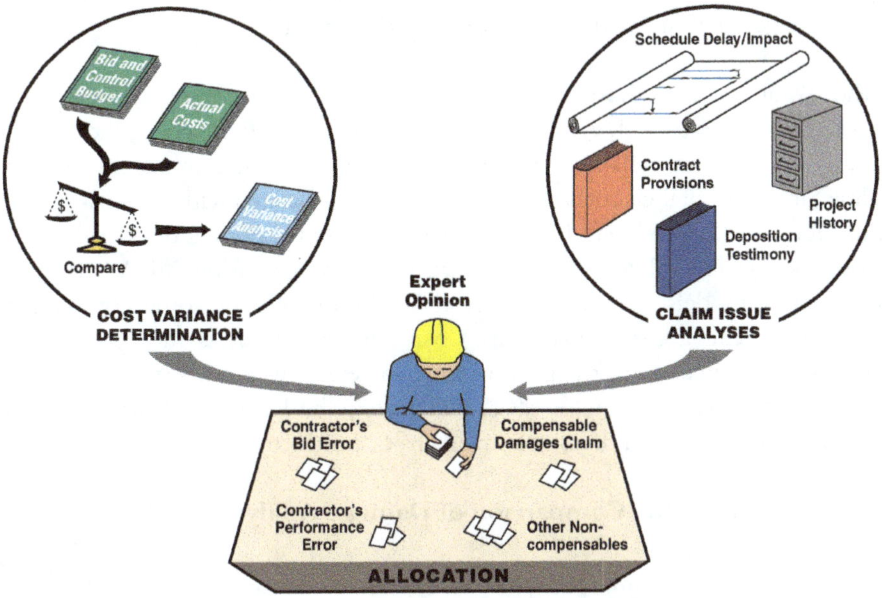

The mathematical structure of the allocation should demonstrate clearly that all variances have been included and totaled to an amount equal to the difference between the contract price and the actual costs incurred. An example of the overall allocation of cost variances into a cost/damages matrix using the Discrete Damages/Cost Variance Analysis method is previously shown by Figure 5-8.[417]

5.2.8 Comparison of Methods

The various methods of calculating damages analyze either part or all of the costs comprising the in-place value of the project. A Total Cost approach will merely request the difference between the bid price, with possible adjustments for approved and pending change orders, and the in-place value, which is defined as the total field costs plus overhead and profit. The "A"/"B" estimate approach tries to evaluate the claim value from independent estimates of damages, hopefully falling between the adjusted value (in-place value less non-compensable costs) and the adjusted bid (bid

417 See Chapter 5, Section 5.2.7.1 herein.

plus bid error). The Jury Verdict approach and the Modified Total Cost approach deduct estimates of both bid errors and non-compensable costs from project costs, leaving a lump sum claim amount. Delta Estimates divide the claim in pieces, with separate costs for each compensable problem, without showing how each claim fits into the total actual costs for the project. Again, the problem is: does the sum of the parts exceed the value of the whole? The Discrete Damages/Cost Variance Analysis method analyzes the entire cost spectrum. All components of bid error, non-compensable costs, and compensable claims are addressed. A comparison of these various damages analysis methods is shown by Figure 5-12.

For the Total Cost method, the Modified Total Cost method, the Jury Verdict method, and the *Quantum Meruit* method, all loss of productivity costs, including costs associated with the cumulative impact of changes and other impacting events, are subsumed into the total claimed costs. Thus, costs associated with cumulative impacts are not segregated. The Delta Estimate method would include a direct calculation of the costs being claimed for cumulative impact, typically using MCAA factors, the Leonard Study models, the CII studies, and Ibbs models, or some combination of these studies. The Discrete Damages/Cost Variance Analysis method can include either a direct calculation of productivity losses using industry studies, or the Measured Mile, Earned Value, or MCAA factors, or an indirect calculation where the contractor claims a portion of its cost overrun for productivity losses.

Figure 5-12: Comparison of Damage Analysis Methods

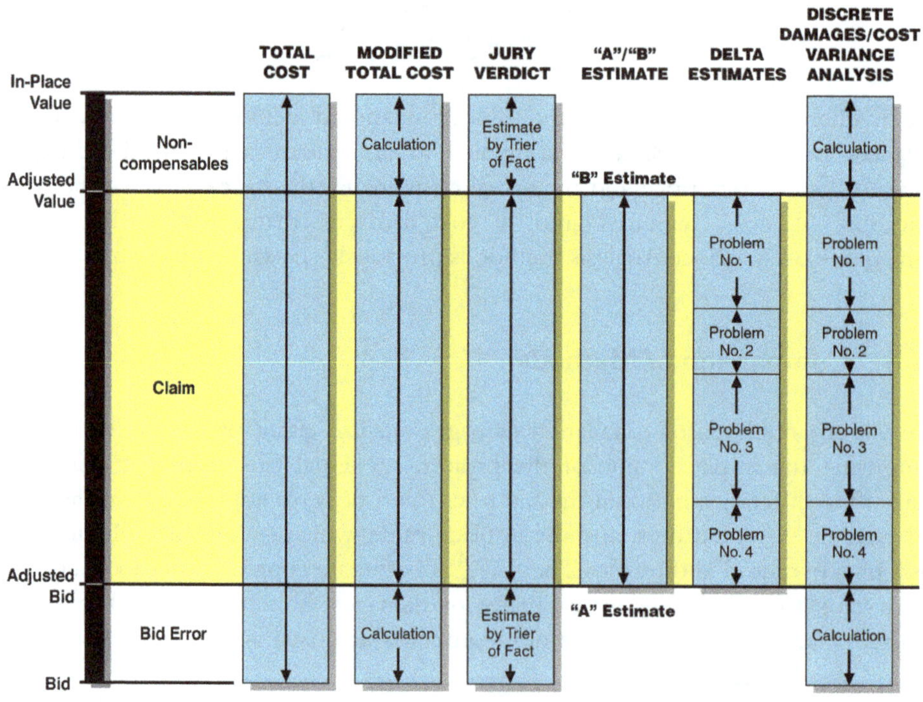

Chapter 6

Supporting a Cumulative Impact or Disruption Claim during the Project

Chapter 6

Supporting a Cumulative Impact or Disruption Claim during the Project

Contractors know when a project is troubled by the impact of many changes and other disruptive events. Upon that realization, if the contractor does not already have adequate systems in place, the contractor should step up its efforts to document the impact to its work and, to the extent possible, discretely quantify the costs of changes and other impacts. Yet when projects are impacted, the opposite often holds true. Foreman and supervisors spend their time coordinating the changed work and finding the most productive work for their crews in an attempt to stay on budget and on schedule; they often have less time to document the impacts to their work and fill out timesheets showing who did what. Depending upon the magnitude of the impacts and potential losses, the contractor should determine whether additional management personnel are necessary to document the impacts to its work.

In this Chapter 6, the following topics are discussed:

- Foreseeability and Reasonable Quantification of Direct Disruption and Cumulative Impact Costs During the Project (see Section 6.1);

- Owner-Directed Change Example (see Section 6.2);

- Options for Contemporaneously Estimating Cumulative Impact (see Section 6.3);

- Documents that a Contractor Should Maintain (see Section 6.4);

- Documenting Changes and Impacts (see Section 6.5);

- Tracking Productivity (see Section 6.6); and

- Notice (see Section 6.7).

6.1 Foreseeability and Reasonable Quantification of Direct Disruption and Cumulative Impact Costs during the Project

The contractor's obligation to include direct disruption and cumulative impact costs (assuming that they exist) in its change order pricing is largely driven by the specific contract provisions and/or the waiver language included on the change order forms. These authors have seen an increased prevalence of language in contracts and change order forms that require the contractor to include all costs, including direct disruption and cumulative impact costs, in its change order pricing, and to waive its rights to claim additional costs at a later date. From the owner's perspective, such language ensures that a change order is fully and finally settled at execution, and that the contractor cannot revisit the change order to submit a later claim for disruption and cumulative impact. From the contractor's perspective, however, if it is clear that significant disruption and cumulative impact costs may result from the change, such language forces it to either: (i) negotiate modified language so that it does not waive its rights to subsequent disruption and cumulative impact costs, a process which may put at risk any payment for the change (*i.e.*, it may become a disputed change), or (ii) attempt to estimate the disruption and cumulative impact costs, which may not be accurate and which the owner will likely view with skepticism.

Generally, there is no debate in the industry that it is mutually beneficial for the owner and the contractor to agree to terms of the change order at the time it is introduced. For example, in regard to good practice for variations or change orders, the Society of Construction Law recommends the following:

> *Where practicable, the total likely effect of variations should be pre-agreed between the Employer/CA and the Contractor, to arrive if possible at a fixed price of a variation, to include not only the direct costs (labour, plant and materials) but also the time-related costs, an agreed EOT and the necessary revisions to the programme...*[418]
>
> *It is not good practice to leave to be compensated separately at the end of the contract the prolongation and disruption element of a number of different variations and/or changes. This is likely to result in the Contractor presenting a global claim, which is a practice that is to be discouraged. Where it is not practicable to agree in advance the amounts for prolongation and disruption to be included in variations and sums for changed circumstances, then it is recommended that the parties to the contract do their best to agree the total amount payable as the consequence of the variations and/or changes separately as soon as possible after the variations are completed.*[419]

---

418 "The Society of Construction Law Delay and Disruption Protocol," Oxford, October 2002, p. 8.

419 *Id.* at Section 1.7.7, p. 20.

These are sound recommendations, but in cases where the contractor attempts to estimate cumulative impact contemporaneously, there are few good options which may lead to agreement between the parties. Below is an example for discussion, as well as potential options for estimating cumulative impact contemporaneously.

6.2 Owner-Directed Change Example

Let us consider an example of an owner-directed change to a contractor's scope of work on an industrial plant expansion project, wherein the owner requests a lump sum change proposal for replacing 1,000 feet of existing piping. The owner further requests that the contractor implement overtime measures to complete the changed work, and both parties are aware that electrical and instrumentation work is ongoing in close proximity to the piping replacement. Per the contract and the change order form language, the contractor is required to include all its costs, including those caused by direct disruption and cumulative impact.

In preparing its proposal, the contractor follows its standard procedure and estimates its direct costs using unit rates. To quantify the costs of overtime, it uses industry studies to estimate its expected labor inefficiency costs, and also includes appropriate increases for its overtime labor rates. The contractor applies a further 10 percent increase to its labor hour estimates of both piping and electrical and instrumentation work (in that area) for expected loss of productivity (direct disruption) due to trade stacking and congestion. The contractor has therefore estimated the costs of its direct extra work and the local or direct disruption that it foresaw at the time. While the owner may take exception to the contractor's disruption factors, and negotiate the proposed costs accordingly, it will likely acknowledge that the contractor will expend additional man-hours due to overtime, trade stacking, and congestion.

But what about the contractor's costs associated with cumulative impact, the compounding effect of this change and others on the contractor's base scope of work and its overall labor productivity? Are these impacts foreseeable, and if so, can the costs of such impacts be accurately estimated at the time the change is being priced?

Regarding the issue of foreseeable impacts, timing matters. If this particular change is one of the first on the project, the contractor most likely has not yet suffered any cumulative impact and cannot foresee potential subsequent changes that may, in combination with piping replacement work, ultimately impact its productivity. If several dozen changes have predated this change, and significant changed work is ongoing, the contractor may and probably should recognize that another change will contribute to its reduced productivity, *i.e.*, it will incur cumulative impact costs.

Generally, while the cumulative impact of events may not be foreseeable early in a project, at some point in time an experienced contractor will or should recognize that it has encountered a sufficient number of changes or impacting events to its work, such that its labor productivity has degraded beyond its reasonable expectations and experience. It is at this point in time that an experienced contractor should know, or at least evaluate whether, it is incurring increased man-hours and costs due to productivity loss caused by the cumulative impact of such problems.

Courts and boards hold contractors to this standard. For example, in 1981 the *Pittman* Board wrote:

> *Twenty or even ten years ago, an Appellant might justifiably have taken the position that it could not have foreseen any sort of impact costs and that it first became aware of them only after it assessed its performance and realized it had a loss. Appellant in this case clearly does not make this contention with respect to all impact costs; instead, it appears to be saying it foresaw the more routine sort of impact costs but it did not foresee those associated with cumulative impact. We consider the more plausible version to be that Appellant accurately assessed the complete cost of the changes including impact.*[420]

Also, it is important to recognize that experienced contractors maintain detailed accounting and cost control systems wherein they are able to recognize cost increases above their budgeted amounts or degradation in productivity in a timely manner. Certainly, on lump sum or fixed price contracts, such systems should be implemented by the contractor. Productivity measurement and Earned Value systems provide data to the contractor such that, if multiple change orders have occurred, the contractor would be able to recognize this effect from the data. However, even with the appropriate systems in place, contractors often fail to recognize the potential impact of changes and may attribute productivity losses to other factors. Ibbs and McEniry stated:

> *This compounding and negative effect often goes unnoticed until it is too late. It generally becomes apparent only in the later stages of a project when work cannot be completed on time and when labor productivity does not measure up to the anticipated levels.*[421]

This same research team also acknowledges that the unexplained extra hours may be due to the contractor's underestimate or the contractor's own inefficiencies, as follows:

> *The difference between actual hours and estimated hours cannot, however, be explained simply by the number of approved change order hours. The balance represents unexplained*

420 *Pittman Construction Co.*, GSBCA Nos. 4,897, 4,923, 81-1 BCA ¶ 14,847, 1980 WL 2643 at pp. 24–25.

421 Ibbs, William and Gerald McEniry, CFCC PSP, "Evaluating the Cumulative Impact of Changes on Labor Productivity – an Evolving Discussion," Cost Engineering, AACE International, Vol. 50/No. 12, December 2008, pp. 23–29 at p. 23.

> *extra hours which may partly be the result of an underestimate, the contractor's own inefficiencies, or, in the case of a project suffering from a multitude of changes, the loss may be the impact of change orders.*[422]

Returning to our example, let us assume that at the time of the proposed change, the project is already highly impacted and the contractor recognizes that its labor productivity is being negatively impacted, at least in part, by the effect of multiple change orders. While cumulative impact of this change (and others) is now foreseeable, the contractor most likely cannot accurately quantify its additional costs, especially in respect of this change alone. As discussed in Chapter 3, Section 3.4.1.3, the contractor's options for estimating cumulative impact are limited, and the studies generally correlate overall productivity loss to the amount of change *at project completion.*

6.3 Options for Contemporaneously Estimating Cumulative Impact

Yet, the contractor can attempt to estimate cumulative impact in the above scenario, even if the accuracy is questionable. The following options assume that there is no contractual mechanism in place to estimate direct or cumulative disruption, such as the NAVSEA Guidelines once used in certain U.S. Navy contracts.[423]

One option may be for the contractor to utilize MCAA factors. Contractors use MCAA factors to forward price loss of labor productivity,[424] and as discussed previously,[425] certain factors are consistent with the theory of cumulative impact. However, the contractor's cumulative impact estimate for the change at hand would likely take the form of a factor applied to the remaining planned labor hours for the entire project, even though the change may directly impact only a small portion of the project. The owner may rightly question the accuracy of such an estimate.

A second option, and likely a more viable one, is for the contractor to utilize one of the several studies (e.g., 1995 CII or 2005 Ibbs) that correlate the amount of project change and overall labor productivity. Under this scenario, the contractor would keep a running total of the additional man-hours associated with changed work, and would therefore be able to quantify the percent of change on the project and the theoretical loss of productivity caused by the changes. Based on the studies,[426] productivity degrades as more changes are introduced. Moreover, the incremental productivity

422 *Id.*

423 See Chapter 3, Section 3.4.1.4.2, for a discussion of the NAVSEA guidelines.

424 See "Change Orders Productivity Overtime – A Primer for the Construction Industry," MCAA, 2012, pp. 81, 86–92.

425 See Chapter 3, Section 3.4.1.4.1.4.

426 See Chapter 3, Section 3.4.1.3.

loss can be estimated. For example, if a given change order increases the overall percentage of changed work from 10 percent to 12 percent, the incremental two percent increase yields an incremental productivity loss that can be estimated and used to determine the theoretical lost man-hours due to cumulative impact.

A third option for the contractor may be an approach discussed by Mr. Pickavance, who contends that loss of efficiency can be evaluated in advance of performing the changed work.[427] He outlines an approach that enables the analysis of local disruption and the ripple effect of the change.[428] The approach assumes that the contractor has implemented a resourced-loaded[429] planned schedule (program). To summarize, the steps involved include: (i) statusing the schedule activities and resources immediately prior to the introduction of the change; (ii) identifying the planned activities and resources that will be impacted by the change; (iii) ascertaining the activities and resources required to implement the changed work, thus establishing a revised program after the change is introduced; and (iv) comparing the "before" and "after" scenarios to value the change.[430]

The approach discussed by Pickavance is a valid one, as showing the impact to schedule activities and resources helps establish the cause-effect relationship (which is discussed further in Chapter 7, Section 7.2). Additionally, the contractor may be required to provide such an analysis in support of a request for a time extension. Thus, to the extent a contractor can prepare such analyses during the project, it is encouraged (if not required). However, impacting the planned schedule/program activities and resources does not, in and of itself, accomplish the task of quantifying man-hour growth. That is, a resource-loaded schedule will not quantify disruption on its own; it is driven by user-defined inputs. The estimation of man-hour growth for a specific activity or group of activities is therefore still reliant on contractor-input assumptions and disruption factors, which must be proven to the owner's satisfaction.

In implementing one or a combination of the above options, the contractor can at least attempt to estimate cumulative impact at the time a change is introduced, which is a necessary step in negotiating and agreeing to a price. Barring contemporaneous resolution, the contractor's best option during the project may be, when the magnitude of change orders reaches a certain percentage of the contract, to inform the

---

427 See Pickavance, Keith, *Delay and Disruption in Construction Contracts,* 3rd ed., T&F Informa (UK) Ltd, London, 2005, p. 665, paragraph 17.31.

428 *Id.* at paragraph 17.33.

429 A resource-loaded schedule is a schedule in which the contractor has assigned manpower or man-hours to specific schedule activities, thus showing the contractor's planned manpower or man-hours for specific scopes of work over time.

430 See Pickavance, Keith, *Delay and Disruption in Construction Contracts,* 3rd ed., T&F Informa (UK) Ltd, London, 2005, p. 665, paragraphs 17.32–17.33.

owner of potential productivity impacts and request a change to the contract specifically for cumulative impact.

6.4 Documents that a Contractor Should Maintain

Document control systems that are employed by large engineering and construction contractors include procedures for storing and retrieving original paperwork. Incoming and outgoing documents are captured, copied, categorized, and filed. Selection of the appropriate capture methods and filing and retrieval systems depends largely on the contractual requirements.

Adequate documentation is vitally important to the contractor interested in survival. First, it provides the information needed to effectively bid, plan, and construct the work. Second, it provides the hard data needed to analyze and successfully resolve the construction claim. A presumption of reasonableness arises in favor of such records that satisfies the claimant's burden and shifts it to the other party, who must then prove that the records are unreasonable and/or inaccurate.

Project records should, at a minimum, be (i) made contemporaneously with the event, (ii) made by, or be based upon, information from a person with knowledge of the event, (iii) kept as a regular practice of the company, and (iv) sufficiently detailed to permit a third person to reconstruct project activities solely from the files.

It is also imperative for contractors and subcontractors to be problem oriented, as well as cost oriented, while maintaining documentation during construction, to be able to identify and react to those problems which may impact the contract price or performance time. It is imperative that records are kept of anything and everything that may be pertinent to problems or claims.

Proving the actual costs lost is critical to a construction claim. Use and maintenance of effective accounting methods can provide the proof of damages necessary to support additional compensation.

For example, the United States Court of Appeals has stated its requirement that the contractor provide its actual costs for the disruption caused by the change order:

> *[T]he issuance of a change order request should signal to the prudent contractor that it must maintain records detailing any additional work ... because the record does not show that [the contractor] justifiably could not have submitted an "actual cost claim," we remand to the*

claims court for a redetermination of the equitable adjustment due [the contractor] by the "actual cost method."[431]

Because delay, acceleration, and impact claims frequently involve inefficiency and loss of productivity factors that are difficult to segregate under traditional accounting systems, a method of isolating costs not covered by the contract is especially needed. A system that allows for concurrent segregation of unanticipated costs is not only easier and less expensive than the after-the-fact breakdown; it is also more convincing during the trial or arbitral proceedings.

Cost reports are often prepared to study unit costs or production units as the job progresses. Analysis of these cost reports may show where the contractor's estimate was incorrect, or where inefficiencies were caused by the contractor's own mismanagement.

Although cost records and other documentation on each project will vary, Appendix C represents the typical spectrum of pre-contract, contract, project management, contemporaneous records, correspondence, schedule, man-hour, quantity, cost, and other information that should be preserved for potential use in proving damages and establishing cause-effect relationships. It also provides a list of the information that the owner should request from the contractor's records to analyze and defend against a contractor's disruption and cumulative impact claims.

6.5 Documenting Changes and Impacts

Even though it is difficult (or impossible) to contemporaneously quantify disruption and/or cumulative impact in terms of man-hours and costs, the field staff must document the changed conditions which are causing the loss of productivity, such as design interferences, crowded work areas, out-of-sequence work, or stop-and-go work. A consistent record of problems that are documented contemporaneously through daily reports, RFIs, change requests, time sheets, and correspondence greatly facilitates any analysis or expert opinion that can be offered after the project is completed. Absent these records, the contractor will find it difficult to recover on any cost overruns, much less those caused by the cumulative impact of changes.

One of the best ways to track owner-caused impacts is through daily reports. A contractor's daily reports typically provide only general information, such as temperature, the number of men each subcontract had on site, and what work was being performed. During a highly impacted project, the foreman or superintendent should use the daily reports as a vehicle to track changed conditions, owner directions,

---

431 *Dawco Constr., Inc. v. United States,* 930 F.2d 872, 880, 882 (1991) (over ruled on other grounds by *Refiectone, Inc. v. Dalton,* 60 F3d 1572 (Fed. Cir. 1995)).

design interference problems, or any variance from the contract. An entry such as, *"Mike's crew stopped work on pipe spool BG-235 (issued RFI 112) – moved to spool BB-002 per CM – approx. 2 hours lost for each of 6 men"* is much more effective than, *"Pipe crews – 20 people today."*

The contractor should also track the costs of change orders and other impacts in separate cost accounts, apart from the cost codes for base scope work. Change orders are often estimated prior to the performance of the changed work, and often the costs to perform that work are greater than estimated. This overrun might not be due to a poor estimate, but caused by the unforeseeable disruption on the changed work, particularly when the work is performed well after the estimate has been prepared and concurrent with numerous unplanned changes that also have to be integrated into the contractor's work schedule. Tracking the costs of changes separately may allow the contractor to recover extra costs through negotiations, or to alert the owner that productivity is being compromised by the cumulative impact of changes.

The benefit to the contractor of discretely identifying the occurrences and magnitudes of changes and impacts during the project is twofold. First, as a general rule, discretely identified impacts that can be quantified (such as the man-hours associated with idle time) are much more likely to be recovered than if the same impacts were grouped into an after-the-fact disruption claim. Second, the identification of impacts, even if they cannot be quantified at the time, can be used to demonstrate causation, and will increase the likelihood of the contractor's recovery on a disruption claim.

Another consideration for the contractor is documenting the amount of time spent by management staff handling changed work. This staff – project managers, project engineers, site managers, schedulers, etc. – typically code their time to general management cost accounts. On highly impacted projects, however, this staff may be spending considerable time preparing and negotiating change order proposals, corresponding with subcontractors and the owner, updating schedules, planning work-arounds, and other tasks associated with managing unplanned work. While the costs of this time may not be recoverable (as management costs are typically driven by project duration and not scope of work), if the time is excessive, it may be proof of a changed project environment that helps explain an overall degradation of productivity. In addition, such records would support any decision to assign additional management staff to the project, the costs of which may be directly recoverable.

6.6 Tracking Productivity

As discussed in Chapter 3, Section 3.4.1.1, the most acceptable method for proving and pricing disruption is the Measured Mile analysis, followed by the Earned Value analysis, which is discussed in Section 3.4.1.2. To effectively utilize either analysis, the contractor must track its installed quantities and actual labor man-hours – both in terms of timing and scope of work (discipline, area, WBS element, cost account, etc.) – such that its actual productivities can be ascertained. Moreover, the productivity data in most cases must be sufficiently detailed such that productivity losses can be ascribed to the alleged impacts suffered by the contractor.

It is recommended that the contractor implement detailed productivity measurements during the project, both to have timely information to identify productivity losses as they occur, and to support change requests or potential claims. If it is not the contractor's standard practice to track productivity at a detailed level, and the contractor finds itself in a highly impacted project, the contractor should, at a minimum, begin tracking installed quantities and actual man-hours for those disciplines or areas that it believes will suffer the most impact.

6.7 Notice

Once the contractor becomes aware that a multitude of changes are impacting its productivity, *i.e.*, the impacts are foreseeable, it is essential that the contractor notify the owner of its findings, and that the contractor reserves its rights to claim such impacts on any subsequent change orders at the end of the project.

Most construction contracts require written notice for changes, differing site conditions, extra work, or other events which may affect the contractor's time and cost of performance. The process of giving "notice" is vital to triggering the contract mechanisms that allow the contractor to pursue additional time and cost, and to reserve its rights to recover for any unforeseen but related impacts that may later arise.

As discussed in Chapter 2, Section 2.2.2, courts and boards have denied many cumulative impact claims in part because the contractor has not taken such steps. Thus, failure to provide proper and timely notice may limit the contractor's ability to recover its increased costs of performance.

Some of the key factors required by standard contract form notice provisions are as follows:

1. Notice must be in writing;
2. Written notice must be served to a specific designated party;
3. Notice must be given as soon as the notifying party becomes aware of the circumstances of an impact event or condition;
4. Notice must be given within a defined time period;
5. Notice for time and additional cost impacts must be given if both are deemed involved;
6. Details of the actual or anticipated effects of any impacts, if practicable;
7. Estimates of the extent of any time and cost impact;

8. Supporting documentation must be submitted within a specified time frame (possible second notice period);
9. Specific schedule and work activities affected; and
10. Cause, effect, and length.

All of the contractually required information associated with a notice, as listed above, may not be immediately available when the event triggering a requirement for a notice occurs. It may not be physically possible for the contractor to understand and evaluate every potential impact event that may occur under a short timeline contractual notice requirement. Often what appears to be a problem on a "first look" basis may resolve itself, or may never have been a factor for consideration of notice in the first place. Conversely, an impact event of merit may not manifest itself immediately in terms of being recognized as a potential claim event requiring notice. A timely notice should still be provided regarding an event or problem, with an indication that more detailed information will be provided at a later date.

Chapter 7

Proving the Cause-Effect Linkage

Chapter 7

Proving the Cause-Effect Linkage

Contractor's claim submittals and expert reports are often deficient in proving causation, *i.e.*, the cause-effect linkage. These claims generally outline the owner-caused impacts and separately calculate quantum; however, the two are often not linked in any meaningful way. The large majority of claims are settled prior to a decision by a panel, court, or board, and therefore these deficiencies are not made apparent. Yet, a well-prepared claim document which includes a persuasive and accurate cause-effect analysis can greatly improve the contractor's chances of a successful recovery, either through negotiations or in arbitration/litigation. This analysis is difficult and often costly to prepare, and is therefore not performed in many disputes, which may be the reason why the claims fail.

For the analyst seeking to show the cause-effect linkage in a cumulative impact of changes claim, the task is even more difficult. By an earlier definition, cumulative impact is *"exclusive of that local disruption that can be ascribed to a specific change,"*[432] or stated another way, cumulative impact cannot be ascribed to a specific change. Yet court and board decisions have shown that simply demonstrating that a large number of changes existed, and that the contractor suffered a loss of productivity, is not sufficient for recovery.[433] This dichotomy is difficult to overcome.

In *Luria Bros. & Co. v. United States*, the court stated: *"It is a rare case where loss of productivity can be proven by books and records; almost always it has to be proven by the opinion of expert witnesses."*[434] Proving causation in direct disruption or cumulative disruption claims most often requires a qualified expert, as well as detailed and complete contemporaneous project documentation. The following suggestions, while case sensitive and highly dependent on the detail in the project record, provide several ideas on linking excessive changes with a loss of productivity:

432 *Centex Bateson Constr. Co.*, VABCA Nos. 4613, 5162, 5165, 99-1 BCA ¶ 30,153 (1998), aff'd, *Centex Bateson Constr. Co. v. West*, 250 F.3rd 761 (Fed Cir. 2000) at 149,258 (citing *Triple "A" South*, 94-3 BCA ¶ 27,194, ASBCA No. 46,866 (1994) at 135,523).

433 See Chapter 2, Section 2.2.1 herein.

434 *Luria Bros. & Co. v. U.S.*, 177 Ct. Cl. 676, 369 F.2d 701, 713 (1966) at 713.

- Prepare a schedule analysis (see Section 7.1);
- Track the impact on an activity or crew (see Section 7.2);
- Show how the site environment changed from plan (see Section 7.3);
- Tell the factual story with graphics (see Section 7.4); and
- Prepare a cause-effect matrix (see Section 7.5).

7.1 Schedule Analyses

Contractors often utilize a schedule analysis to identify and help quantify the effects of owner-caused changes and impacts that delayed their work. As a result of project delay, the contractor may accelerate, increase its manpower, work excessive overtime, experience trade stacking and crowded conditions, and allege compensable loss of labor productivity. However, without support from a schedule analysis, the contractor makes a quantum leap from proving the owner-caused delays and impacts, to alleging that 100 percent of the productivity loss is compensable due solely to the owner-caused delays and impacts.

The contractor's own problems may have contributed to the productivity loss and must be considered. A more appropriate analysis of the increased labor costs would be to allocate the labor cost increases to both the owner and the contractor according to their respective share of delays, disruptions, and problems.

Contractors are not legally bound to demonstrate delayed performance or project delay to recover for loss of productivity.[435] However, the relationship between delay and productivity is evident. If a contractor is delayed, it may have to recover that delay through acceleration measures, at the expense of productivity. Likewise, if a contractor suffers poor productivity, it may lead to schedule delay.

When courts and boards are considering causation in respect of disruption claims, the proof of delay, or lack thereof, is at least one consideration. In *Charles G. Williams Construction*, the board found that the contractor had proven it was delayed unreasonably by the government's actions, and therefore the contractor was entitled to damages for both extended overhead and disruption.[436] In discussing proof of causation in *J. A. Jones*, the board emphasized that the contractor had not proven it was delayed as a result of compensable issues, which was one basis for its denying the contractor's cumulative impact claim.[437] In *Advanced Engineering & Planning Corp.*, the

---

435 See "Estimating Lost Labor Productivity In Construction Claims," AACE International Recommended Practice No. 25R-03, April 13, 2004, p. 6.

436 See *Charles G. Williams Constr., Inc.*, ASBCA No. 33766, 89-2 BCA ¶ 21,733 (1989).

437 See *J. A. Jones Construction Co.*, 00-2 BCA ¶ 31,000, ENGBCA Nos. 6390-1, 6386-7, 6348, 6388-9, 2000 WL 1014011 at p. 48.

board, in denying the contractor's disruption claim for lack of causation, cited the contractor's failure to use a resource-loaded CPM schedule as required by contract:

> *In this connection, we have found, had it submitted and updated a "resource loaded" CPM as required by Standard Item No. 009-60, AEPCO could have tracked disruptive impact through the schedule...*[438]

Based upon the above, it is evident that a schedule analysis can be helpful in demonstrating causation. Even if a formal schedule analysis is not prepared, bar charts, histograms, and other data sourced from the project schedule can be useful in demonstrating causation. Several examples of this are provided in this chapter.

7.2 Tracking Impacts by Activity or Crew

One definition of cumulative impact that was cited earlier is when work on one activity is adversely affected by another activity, or by the mere nature of the site environment.[439]

In a claim submittal, it can be effective to demonstrate the impacts affecting one activity or a group of activities. After that impact is established, one may be able to show the "ripple," or the impacted activities' negative effect on other activities. For example, with detailed project documentation, the following description could be provided:

> *The work on pipeline 2A01 was planned to take 10 work days, but it actually took 30 work days. In that period, five RFIs were issued requesting clarification on design discrepancies. The owner took on average four days to respond. The owner issued Rev 2 isometric drawings on September 3rd, ten days after work had commenced, which led to rework. Further, owner-supplied valves were late and much of the pipe could not be installed until the valves were set. As shown in the daily reports, there were constant interruptions to the work...*

From the project schedules, one could then determine which activities ran concurrently with this activity, and which activities were its logical successors. The changes on pipeline 2A01 may have impacted these successor activities as well. If the above description of the impacts to pipeline 2A01 included "work delayed by open excavation of under-ground pipe for 2B18," the theoretical "ripple" becomes real.

---

438 *Advanced Engineering & Planning Corp.*, ASBCA Nos. 53366, 54044, November 2004, at 104.

439 See Robert F. Cushman, Stephen D. Butler, & James F. Nagle, *Construction Change Order Claims*, Wiley Law Publications, § 4.11 (1998 Supplement).

Even though the work described may be a small percentage of the total scope, providing just a few such examples begins to demonstrate cumulative impact. The next step, per the definition above, is showing how the overall site environment was impacted.

7.3 Site Environment Changes

Changes to the site environment can be any difference in the way work was actually performed, compared to the way it was planned to be performed. Both MCAA and Leonard provide typical causes of lost productivity, which could also be considered site environment changes. Leonard provided the following description:

> *Cumulatively, change orders result in the following causes of productivity loss: stop-and-go operations; out-of-sequence work; loss in productive rhythm; demotivation of work force; loss in learning curve; unbalanced crews; excessive manpower fluctuations; unbalancing of successive operations; lack of management and engineering support; and acceleration when equitable time extensions are not granted.*[440]

The cause-effect link is shown by proving that these conditions existed. The contractor's best case can be made if these impacts are recorded contemporaneously, such as in daily reports. Assuming there is no such record, a few examples of analyses that might be performed to demonstrate this proof are discussed below:

Out-of-Sequence Work: A comparison could be made of the contractor's as-planned and as-built schedules, showing which activities were performed out-of-sequence from plan. One could review the causes of this sequence change and show a detailed list of activities that were impacted due to changes.

Using the same comparison, one could review concurrent activities in the planned schedule compared to those in the as-built schedule. If only four piping activities were planned to run concurrently, but the as-built schedule shows that ten piping activities ran concurrently, this may help prove productivity loss. This analysis would correspond to the MCAA factor "Concurrent Operations."[441]

In Figure 7-1, for example, the planned schedule shows the contractor anticipated that piping work would proceed through the four areas with very little concurrent work. Due to late design and site access issues, the work started late and out-of-sequence, and the contractor actually performed work in three or four areas simultaneously.

440 Charles A. Leonard, "The Effects of Change Orders on Productivity," M.S. Thesis at Concordia University, 1988, p. 121.

441 See "Change Orders Productivity Overtime – A Primer for the Construction Industry," MCAA, 2012, p. 77.

Figure 7-1: Bar Chart Schedule and Manpower Comparisons

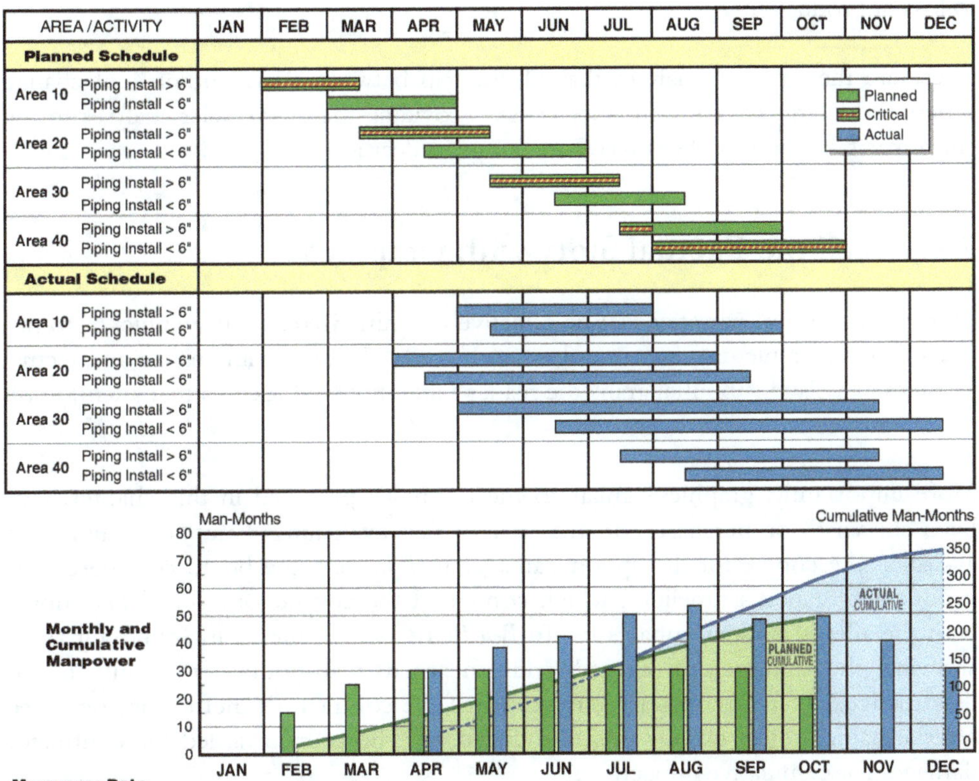

Manpower Data												
Planned Manpower	–	15	25	30	30	30	30	30	30	20	–	–
Cumulative Planned Man-months	–	15	40	70	100	130	160	190	220	240	–	–
Actual Manpower	–	–	–	30	38	42	50	53	48	49	40	30
Cumulative Actual Man-months	–	–	–	30	68	110	160	213	261	310	350	365

<u>Demotivation of Work Force</u>: From human resources and cost records, one could calculate the absenteeism and turnover rates on the project and compare them with other similar projects that were not as severely impacted by changes. If the increased absenteeism and turnover occurred during the same period as did the owner-caused impacts, the cause-effect relationship may be justified.

<u>Loss in Learning Curve</u>: One could make a comparison between the planned peak manpower and the actual peak manpower, and potentially show that the timing of additional forces corresponded to the timing of change order work. Were the added workers craftsmen or laborers, and what was the impact? How much time was lost by new labor crews learning the remaining scope of work, availability and location of tools and equipment, location of materials, project specific construction procedures, etc.? In Figure 7-1, the contractor planned to complete the work utilizing 30 workers,

but actually used 53 workers at the peak. It is likely that the addition of 23 workers led to loss of productivity.

These are the types of analyses that fill the gap between a large number of change orders and a large overrun on labor costs. Providing these or similar analyses to the court may take some of the subjectivity out of its decision.

7.4 Tell the Factual Story with Graphics

Dozens of pages of narrative can be effectively summarized in one graphic, or to put it another way, a picture's worth a thousand words. Including interesting and factual graphics in a claim submittal grabs the reader's attention and focuses that attention on one's argument.

More importantly, graphics similar to the examples provided in this chapter have been shown to be persuasive to the trier of fact evaluating disruption claims. In awarding the contractor disruption damages in *P.J. Dick*, the board considered the productivity expert's timelines, which compared the contractor's actual man-hours with the alleged project impacts.[442] In *Bell BCI Company*, the court found a graphic showing the contractor's earned and unearned man-hours by month to be "instructive," as it demonstrated that most of the contractor's inefficiency occurred after a major project change was introduced.[443] The court awarded the contractor damages for cumulative impact.

In respect of cumulative impact claims, graphics can also be effective in demonstrating the quantity, timing and magnitude of the changes. Figure 7-2, for example, a graphical example of tracking impacts by activity (as discussed in Chapter 7, Section 7.2), shows the quantity and timing of the Design Change Notices that impacted a given scope of work.

Another demonstrative graphic of impacts is shown by Figure 7-3, which plots the location of various types of impacts on the plan view of a hospital building. This "Measles Chart" tells the story that the design problems which disrupted the construction contractor's work were significant, and logically had to have a ripple effect on its labor productivity. The same chart could be reproduced in monthly intervals to demonstrate the timing of the introduced changes, and that work was impacted over a significant period.

---

442 See *P.J. Dick, Inc.*, 01-2 BCA ¶ 31,647, VABCA Nos. 5840-5850, 5951-5959, and 6017-6024 (2001), at pp. 34–35 (Westlaw print).

443 See *Bell BCI Co. v. United States*, No. 03-1613C, 81 Fed. Cl. 617, 2008 U.S. Claims LEXIS 116 (April 2008), at pp. 19–20.

Figure 7-2: Example of Tracking Impacts by Activity

Code	No.	Description	Start Date	End Date	Schedule Activity Affected
CTG1	CTG1	**Total Overall CTG1 Delay: 62 CDs**	**5/14/12**	**7/13/12**	
CTG1	CTG1	As-Planned Performance as of April 3, 2000 Schedule Update (4/3/12 Schedule Status)	4/3/12	5/13/12	
CTG1	CTG1	Window No. 1 Delay – 14 Cal Days (Based on 5/8/12 Schedule Status Update)	5/14/12	5/27/12	
CTG1	CTG1	Window No. 2 Delay – 24 Cal Days (Based on 6/5/12 Schedule Status Update)	5/28/12	6/20/12	
CTG1	CTG1	Window No. 3 Delay – 20 Cal Days (Based on 6/19/12 Schedule Status Update)	6/21/12	7/10/12	
CTG1	CTG1	Window No. 4 Final Delay 4 Cal Days	7/11/12	7/14/12	
CTG2	CTG2	**Total Overall CTG2 Delay: 66 CDs**	**5/17/12**	**7/21/12**	
CTG2	CTG2	As-Planned Performance as of April 3, 2000 Schedule Update (4/3/12 Schedule Status)	4/3/12	5/16/12	
CTG2	CTG2	Window No. 1 Delay – 25 Cal Days (Based on 5/8/12 Schedule Status Update)	5/17/12	6/10/12	
CTG2	CTG2	Window No. 2 Delay – 20 Cal Days (Based on 6/5/12 Schedule Status Update)	6/11/12	6/30/12	
CTG2	CTG2	Window No. 3 Delay – 10 Cal Days (Based on 6/19/12 Schedule Status Update)	7/1/12	7/10/12	
CTG2	CTG2	Window No. 4 Final Delay 11 Cal Days	7/11/12	7/21/12	
		Design Change Notice (DCN) Impacts			
DCN	H3181-117	Add Cable 11BFE1234 Betw MCC & CJT Cabinets	4/6/12	4/26/12	CTG Electrical
DCN	H3181-118	Revise Cable Terminations	4/6/12	4/6/12	CTG Electrical
DCN	H3181-120	Added Cables 11BUB1232/12BUB1232	4/6/12	4/6/12	CTG Electrical
DCN	H3181-120	Install 125 DC to CJT Cabinets	4/6/12	4/27/12	CTG Electrical
DCN	H3181-127	Revised Cable List	4/6/12	4/6/12	CTG Electrical
DCN	H3181-129	Purchase and Install MBY22 Boxes	4/6/12	5/1/12	CTG Electrical
DCN	H3181-118	Revised Cable Terminations 11CJT01	4/7/12	4/27/12	CTG Electrical
DCN	H3181-129	Add H3181-054-X-1122&H3181-054-X-1222	4/7/12	4/7/12	CTG Electrical
DCN	3181-0004	11CHA11 Cabinet List	4/10/12	4/10/12	CTG Electrical
DCN	3181-0005	11CHA11 Cabinet Wire List	4/10/12	4/10/12	CTG Electrical
DCN	414-DCN	Ignition Transformers	4/10/12	4/24/12	CTG Electrical
DCN	H3181-101	Blowoff Valve Proximity Switches	4/10/12	4/29/12	CTG Electrical
DCN	H3181-115	Proximity Switches	4/10/12	4/10/12	CTG Electrical
DCN	H3181-116	Proximity Switches	4/10/12	4/10/12	CTG Electrical
DCN	H3181-121	Proximity Switches	4/10/12	4/10/12	CTG Electrical
DCN	H3181-121	Rewire 482 Card	4/10/12	4/10/12	CTG Electrical
DCN	H3181-122	Delete Pages to AIC & CFA Cable	4/10/12	4/10/12	CTG Electrical
DCN	H3181-125	CJR01 Power Distribution (Replaces #31)	4/10/12	4/26/12	CTG Electrical
DCN	H3181-125	Delete Pages to AIC & CFA Cable	4/10/12	4/10/12	CTG Electrical
DCN	H3181-125	Power Distribution 11/12CJR01	4/10/12	4/10/12	CTG Electrical
DCN	H3181-126	Delete Pages to AIC & CFA Cable	4/10/12	4/10/12	CTG Electrical
DCN	H3181-127	Drawing and Cable Revisions Unit #11	4/10/12	5/1/12	CTG Electrical
DCN	H3181-127	Revised Cable List	4/17/12	4/17/12	CTG Electrical
DCN	H3181-114	Change Signal Cables from 11CJP01 to 12CJP01	4/18/12	4/18/12	CTG Electrical
DCN	H3181-114	Rev. Connections	4/18/12	4/18/12	CTG Electrical
DCN	H3181-128	Drawing and Cable Revisions Unit #12	4/18/12	4/18/12	CTG Electrical
DCN	H3181-100	Connections to DCS need to be revised	4/21/12	4/21/12	CTG Electrical
DCN	H3181-100	Limit Switches on MBP Changed with Proximity Sw.	4/21/12	4/29/12	CTG Electrical
DCN	H3181-102	Connections to the DCS – Both Units Needs Rev	4/21/12	4/21/12	CTG Electrical
DCN	H3181-110	Corrections to Terminal Points	4/21/12	4/21/12	CTG Electrical
DCN	H3181-109	Fiber Optics	3/25/12	4/23/12	CTG Electrical
DCN	H3181-103	Diesel Generator DCS Terminations	4/27/12	4/27/12	CTG Electrical
DCN	H3181-149	Heat Trace Controllers & Wiring #11 and #12	5/19/12	5/25/12	CTG Electrical
DCN	H3181-107	Evaporative Coolers Wiring Unit #1	4/21/12	5/3/12	Evaporator
DCN	H3181-108	Evaporative Coolers Wiring Unit #2	4/22/12	5/24/12	Evaporator
DCN	H3181-152	Evap. Cooler OM Screens	6/1/12	6/1/12	Evaporator
DCN	H3181-102	Burner Temperature Protection	4/21/12	4/21/12	FM200-Fire Protect
DCN	H3181-111	Reloc Fire Protect Alarm From 11CJP01 to 12CJP01	4/21/12	5/1/12	FM200-Fire Protect
DCN	H3181-110	Fire Panel Terminations	5/8/12	5/24/12	FM200-Fire Protect
DCN	H3181-110	Fire Panel Terminations	4/24/12	4/24/12	FM200-Fire Protect
DCN	H3181-123	Fuel Gas Control Valves	4/10/12	4/26/12	Fuel Gas
DCN	H3181-119	Hydrogen Cooled Generator Junction Box Dwgs	4/10/12	5/4/12	H/CO
DCN	H3181-144	H2 Discharge Operator Modifications	5/16/12	5/16/12	H/CO
DCN	H3181-150	H2 Skid Transmitters	5/20/12	5/25/12	H/CO
DCN	H3181-151	Hydrogen Purge Pushbutton Circuit	5/20/12	5/20/12	H/CO
DCN	H3181-115	Replace 1A Fuses w 2A Fuses in Unit 1 & 2CCJP41	4/10/12	5/18/12	Startup & Commiss
DCN	H3181-130	Remote Load Control Capability	5/19/12	5/19/12	Startup & Commiss
DCN	H3181-130	Remote Load Control	5/23/12	5/23/12	Startup & Commiss
DCN	H3181-116	Check Out I/O Points in Cabinet CJP01	4/10/12	5/2/12	Teleperm
DCN	H3181-126	Flight Recorder	4/10/12	4/29/12	Teleperm
DCN	H3181-111	BOP I/O Points/Fire Protection Alarm	4/21/12	4/21/12	Teleperm
DCN	H3181-135	Software Changes for Fuel Gas Supply (BOP)	5/8/12	5/9/12	Teleperm
DCN	H3181-136	Software Changes	5/8/12	5/8/12	Teleperm
DCN	H3181-137	Software Changes	5/10/12	5/10/12	Teleperm
DCN	H3181-138	Software Changes	5/9/12	5/9/12	Teleperm
DCN	H3181-139	Software Changes	5/10/12	5/10/12	Teleperm
DCN	H3181-140	Software Changes	5/10/12	5/10/12	Teleperm
DCN	H3181-141	Software Changes	5/22/12	5/25/12	Teleperm
DCN	H3181-146	11/12 MBP31, 32& 33AH001 Software Changes	5/23/12	5/23/12	Teleperm
DCN	H3181-112	Add Vibro-Meter Electronic Rack to 11CJQ Cabinet	4/21/12	5/2/12	Vibrometer
DCN	H3181-113	Add Vibro-Meter Electronic Rack to 12CJQ Cabinet	4/25/12	5/24/12	Vibrometer
DCN	H3181-119	Gen. MK Units 11 & 12	4/10/12	4/10/12	Other
DCN	H3181-121	Compressor Blowoff MBA 10AT001 (11 & 12)	4/10/12	5/19/12	Other
DCN	H3181-122	Delete AIC System	4/10/12	4/29/12	Other

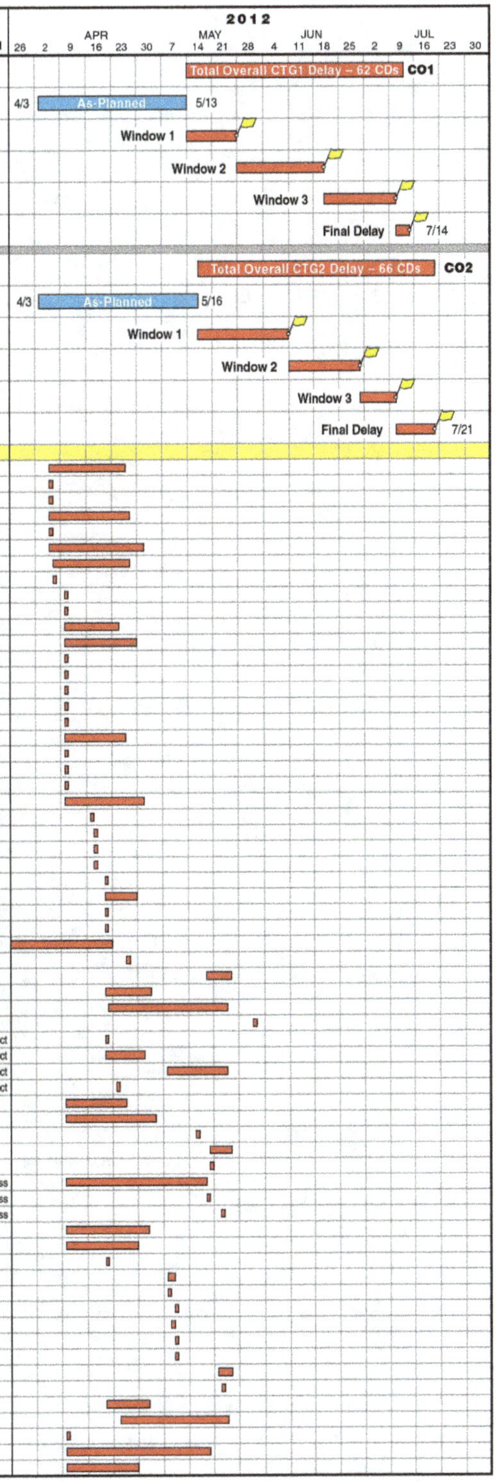

Figure 7-3: Multiple Impacts to Project

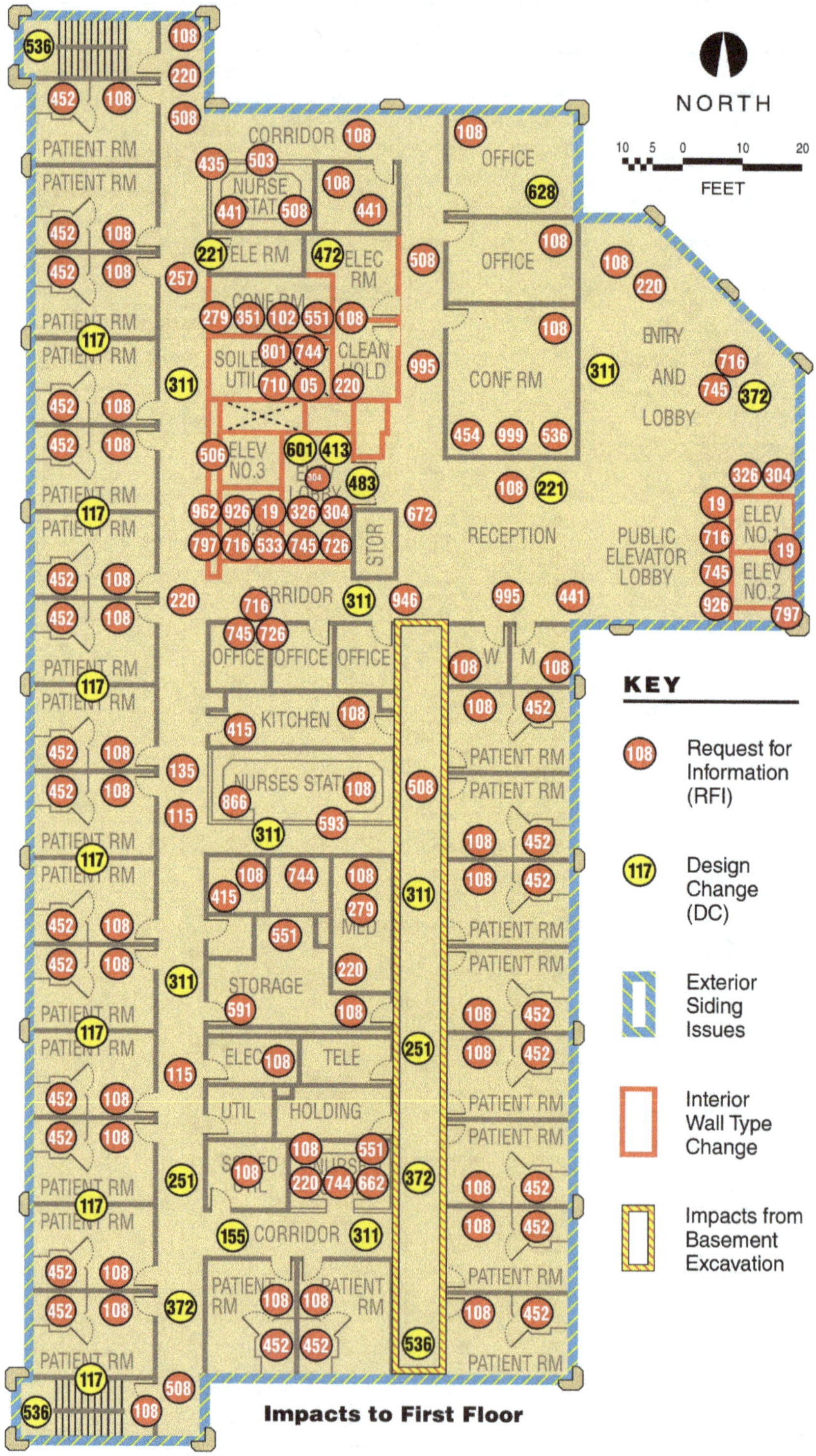

Impacts to First Floor

7.5 The Cause-Effect Matrix

The backbone of the cause-effect linkage in any claim is the cause-effect matrix. For any given owner-caused problem, one can graphically trace the effects to the contractor's work. Figure 7-4 displays simple cause-effect matrices. The first matrix shows that a single problem, defective specifications, results in hydrostatic test failures. The second matrix has three initial causes, two intermediate effects, and two resulting problems.

Figure 7-4: Simple Cause-Effect Matrices

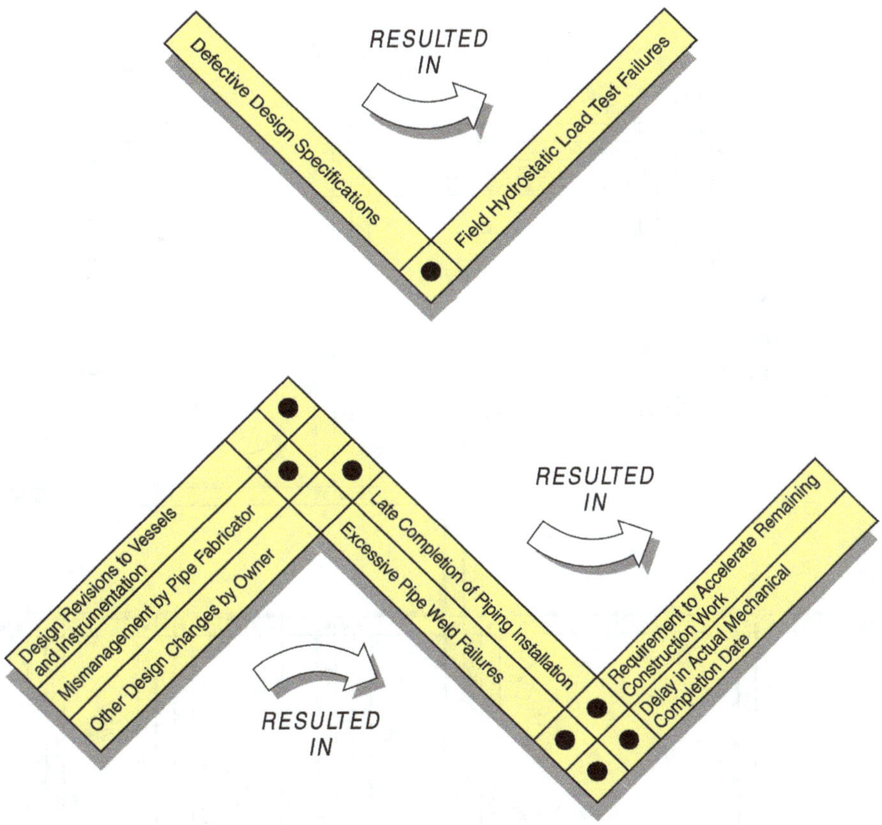

As multiple causes and their resultant effects are added, the matrix ultimately becomes much more complicated. Figure 7-5 shows a typical cause-effect matrix for a highly impacted project. Primary and secondary causes, including contractor-caused problems, are shown to have multiple and duplicative effects, with the end result being a cost overrun. Even though it is complex, Figure 7-5 can still be broken down into the same types of simple relationships that are shown in Figure 7-4.

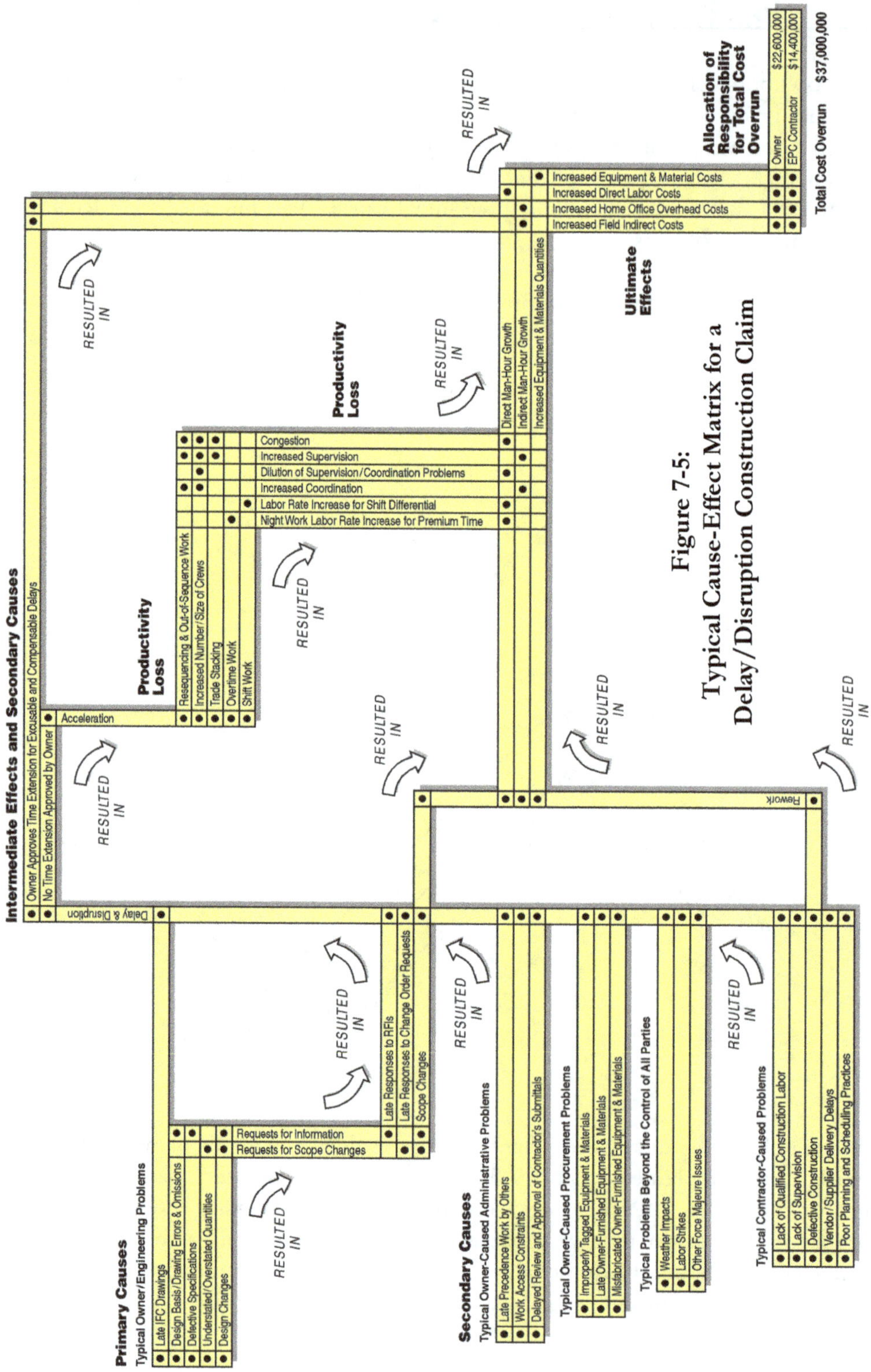

Figure 7-5:
Typical Cause-Effect Matrix for a Delay/Disruption Construction Claim

Tracing one path through this matrix tells a factual story: design changes resulted in the contractor requesting scope changes; the owner was late in responding to many change order requests, which resulted in delay and disruption; the owner did not approve the proper time extensions, which resulted in the contractor accelerating its work; and the acceleration caused trade stacking and congestion, which led to direct man-hour growth and increased direct labor costs, a portion of which are being claimed against the owner. Each of the causes has a similar factual story, the ultimate effect being increased costs.

Showing these relationships graphically, in addition to a narrative containing relevant excerpts from contemporaneous documents, provides the contractor, the owner, the attorneys, and the arbitration panel or court a better understanding of all of the impacts on the project. A similar but more detailed graphic could be prepared showing the effects of several change orders during a given time period to help support a claim for disruption.

Appendix A

Court/Board Decisions Awarding the Contractor Impact Costs

The 18 cases below include court/board decisions in which the contractor was awarded costs for impact claims, even when its claimed costs were significantly reduced. In six cases (Nos. A1–A6), the court/board specifically discussed the theory of, and/or precedence in regard to, cumulative impact. In twelve cases (Nos. A7–A18), the court/board addressed claims for the disruption allegedly caused by multiple changes or impacts, but (i) the term "cumulative impact" was not specifically referenced or (ii) the theory/precedence of cumulative impact was not specifically discussed.

No.	Case	Citation	Project/ (Discipline)	Damages Methodology	Notes on the Disruption Portion of Claim
A1	*David H. Tierney Jr.*	GSBCA Nos. 7107, 6198, 5585, 88-2 BCA ¶ 20,806 (1988).	Office Building	Total Cost; alternatively, a series of allotted values for jury verdict assessment	Contractor claimed for cumulative impact associated with 44 COs (133 separate items). Board found that cumulative impact was not considered in negotiations or CO pricing. Board rejected total cost and contractor's allotted values, but under jury verdict awarded contractor 40 percent of its real project losses, deemed a "fair and reasonable approximation."
A2	*Bechtel National*	NASA BCA No. 1186-7, 90-1 BCA ¶ 22,549 (1989).	Modification of Mobile Launch Platform for NASA	Modified Total Cost (Board deemed it "Total Cost," although it is clear that contractor made adjustments)	Cumulative impact claim for multiple COs, defective drawings specs, and defective Government-furnished equipment. Claims based on 1,209 RFIs, of which 848 allegedly caused impact. Board found that RFIs caused disruption, but not to the extent claimed. By jury verdict, awarded $0.46 M of $7.95 M cost overrun.

Appendix A – Court/Board Decisions for the Contractor

No.	Case	Citation	Project/ (Discipline)	Damages Methodology	Notes on the Disruption Portion of Claim
A3	*Hensel Phelps Construction Co.*	GSBCA 14744; 01-1 BCA ¶ 31,249 (2001).	Office/ Laboratory Complex	MCAA Factors	Impact claims associated with multiple changes, scope revisions, lack of timely information, acceleration, and cumulative impact. Board awarded full claim, finding that expert's analysis and testimony was credible. Analysis involved application of MCAA factors to planned man-hours, detailed by building and timeframe.
A4	*Dugan & Meyers Constr. Co., Inc. v. Ohio Dept. of Adm. Servs.*	2003-Ohio-3709 (2003).	Building	Claimed total costs above adjusted bid for "general conditions" accounts	Court of Claims awarded cumulative impact damages for excessive number of design changes (700 RFIs, 250 field work orders, and 85 architectural supplemental instructions). Overturned on appeal (See Appendix B, No. B20)
A5	*Bell BCI Co. v. United States*	No. 03-1613C, 81 Fed. Cl. 617, 2008 U.S. Claims LEXIS 116 (April 2008).	Laboratory Building	Measured Mile (Earned Value)	Claims for cumulative impact based on 206 contract mods (from 730 EWOs) that delayed completion by 19.5 months and increased original contract price by 34 percent. Board found expert's time impact analysis demonstrated delay and quantum analysis (based upon the contractor's detailed productivity data) sufficient proof of damages. Board found "chaotic" environment due to the number of changes coupled with no time extension.
A6	*States Roofing*	ASBCA No. 55505 (January 2010).	Repair of Building Roofs	Modified Total Cost, supported by Measured Mile-type analysis and also MCAA Factors	Disruption and cumulative impact claims associated with changed working conditions and numerous changes. Award based on "obvious impacts," which the Navy did not dispute. The board was not persuaded by any of the quantum analyses set forth by the contractor, but based on jury verdict, it awarded 10 percent of the contractor's total actual labor costs.

Appendix A – Court/Board Decisions for the Contractor

No.	Case	Citation	Project/ (Discipline)	Damages Methodology	Notes on the Disruption Portion of Claim
A7	*Fruehauf Corp.*	PSBCA No. 477, 74-1 BCA ¶ 10,596 (1974).	Install Mail Handling System (Mechanical)	Not specified	Court awarded approximately one-third of claimed costs (jury verdict) due to large volume of changes (103 total) and late changes. Found that release language did not include "impact" costs.
A8	*Ingalls Shipbuilding Division, Litton Systems, Inc.*	ASBCA 17579, 78-1 BCA, ¶ 13,038 (1978).	Construction of Submarines, Naval Surface Vessels, and Commercial Vessels	Modified Total Cost	Large, complex case wherein the shipbuilder made claims for delay, impact, and displacement costs associated with multiple vessels under multiple contracts, based upon significant changes to submarine contract (over 29,000 revisions to plans) and their impact upon other contracts. Board awarded shipbuilder over 90 percent of its claimed impact costs, making certain adjustments to the modified total cost for bid error and other factors.
A9	*Chantilly Construction Corp.*	ASBCA No. 24138, 81-1 BCA ¶ 14,863 (1980).	Buildings and Sitework	Not specified	Contractor sought equitable adjustment to recover impact and delay costs resulting from various formal contract modifications. Board found that there was no "meeting of the minds" on subject of delay/impacts, and contractor could pursue additional costs.
A10	*C. Norman Peterson Co. v. Container Corp of Am.*	218 Cal Rptr. 592 (Cal. Ct. App 1985).	Paper Mill Modernization	Total Cost/ *Quantum Meruit*	Appeals court affirmed trial the court's decision that Owner had abandoned contract due to poor design, errors, and multiple changes. Expended direct costs exceeded those in original contract by 71 percent. Contractor awarded total costs less amounts already reimbursed.

No.	Case	Citation	Project/ (Discipline)	Damages Methodology	Notes on the Disruption Portion of Claim
A11	*State of California ex rel. Dept. of Transportation v. Guy F. Atkinson Co.*	187 Cal. App. 3d 25 [231 Cal. Rptr. 382] (1986).	Highway Construction	Not specified, but appears to be Total Cost	Appeals court affirmed arbitral decision that the contractor was entitled to disruption damages due to COs which altered the character of the original contract. Arbitration panel, in awarding 65 percent of claimed costs, found that there were "cumulative effects of all the ordered changes" and it was not feasible to separately identify impact of each CO.
A12	*Charles G. Williams Constr., Inc.*	ASBCA No. 33766, 89-2 BCA ¶ 21,733 (1989).	Maintain, Repair, Alter Officers' Club	Not specified	26 contract mods that increased original price by 15 percent. Contractor proved delay was caused by changes and late response/funding thereof, and was therefore entitled to extended overhead costs and disruption damages. Term "cumulative impact" is not used specifically.
A13	*Atlas Construction Co. Inc.*	90-2 BCA ¶ 22,812, GSBCA Nos. 8593, 7903, 8143, 8653 (1990).	Courthouse Renovation	Total Cost (with adjustments)	Board found that contractor's performance burdened by the number of COs, hidden conditions, and direct changes. Total claimed costs awarded based (in part) on worker testimony that 50 percent more interference was experienced on this project than is typical.
A14	*Kit-San-Azusa, J. V. v. United States*	32 Fed Cl. 647 (1995).	Pumping Plant	Total Cost associated with piping work	Court found that the number of design changes was "dramatic and exceeded what the contractor reasonably should have expected." Court awarded 20 percent of cost overrun in consideration for other problems not caused by the government.
A15	*Clark Concrete Contractors, Inc. v. General Services Administration*	99-1 BCA ¶ 30280, GSBCA No. 14340 (1999).	Office Building	Measured Mile and MCAA Factors	Board awarded $6.2 MM of $8.1 MM total claim. Awarded disruption damages caused by "massive changes" to design and slow response to submittals and RFIs.

Appendix A – Court/Board Decisions for the Contractor

No.	Case	Citation	Project/ (Discipline)	Damages Methodology	Notes on the Disruption Portion of Claim
A16	*Amelco Elec. v. City of Thousand Oaks*	98 Cal. Rptr. 2d 159 (Cal. Ct. App. 2000).	City Hall/Arts Theatre/ Meeting Complex (Electrical)	Total Cost	Appeals court upheld a jury verdict that awarded a contractor its total costs due to the city's abandonment of the contract due to excessive changes. Court found city demanded excessive changes that expanded scope beyond contemplation of the parties. 248 drawing revisions impacted work; hours in approved COs increased original estimate by 40 percent.
A17	*Clark Construction Group, Inc.*	00-1 BCA ¶ 30,870, VABCA No. 5674 (April 2000).	Building – Hospital (Mechanical)	Measured Mile, MCAA Factors, and Modified Total Cost	Involved multiple heads of claim with differing damages methodologies. Board rejected claims for disruption caused by multiple late RFI responses due to lack of substantiation from documents and analyses. Board granted disruption damages caused by change in construction sequence and wet conditions (used MCAA factors to quantify award).
A18	*P.J. Dick, Inc.*	01-2 BCA ¶ 31,647, VABCA Nos. 5840-5850, 5951-5959, and 6017-6024 (2001).	Addition to a Medical Center	Measured Mile; alternatively, MCAA factors	Claimed loss of productivity experienced by an electrical subcontractor due to severe design deficiencies/changes and delay/acceleration. Board found measured mile analysis reasonable and valid.

Appendix B

Court/Board Decisions Rejecting the Contractor's Impact Claims

The 31 cases below include court/board decisions in which the contractor's impact claims were rejected. In 24 cases (Nos. B1 – B24), the court/board specifically discussed the theory of, and/or precedence in regard to, cumulative impact. In seven cases (Nos. B25 – B31), the court/board addressed claims for the disruption allegedly caused by multiple changes or impacts, but (i) the term "cumulative impact" was not specifically referenced, or (ii) the theory/precedence of cumulative impact was not specifically discussed.

The Reasons for Rejection noted below are as follows:

1 Failure to prove liability (general);
2 Failure to prove liability (lack of notice or reservation of rights);
3 Failure to prove causation;
4 Causation/quantum – specific notes on lack of contemporaneous productivity data and support for impacts;
5 Failure to prove quantum; and
6 Barred – no recognition of cumulative impact

No.	Case	Citation	Project/ (Discipline)	Damages Methodology	Notes on the Disruption Portion of Claim	Reasons for Rejection
B1	*Acme Missiles & Constr. Corp.*	ASBCA Nos. 11256, 11716, 68-1 BCA ¶ 6,873 (1968).	Dormitory	Not specified	Contractor "failed to show any costs resulting from the changes to the work under this contract other than those which it negotiated with the contracting officer and those we have allowed above."	2, 3, 5
B2	*Dyson & Co.*	ASBCA No. 21,673, 78-2 BCA ¶ 13,482, on reconsid., 7901 BCA (CCH) ¶ 13,661 (1979).	Hospital (Mechanical)	Not specified		1, 2
B3	*Pittman Construction Co.*	GSBCA Nos. 4,897, 4,923, 81-1 BCA ¶ 14,847, 1980 WL 2643.	Office Building, Courthouse, and Parking Facility	Claims from Pittman subcontractors; methods not specified		3

Appendix B – Court/Board Decisions not for the Contractor

No.	Case	Citation	Project/ (Discipline)	Damages Methodology	Notes on the Disruption Portion of Claim	Reasons for Rejection
B4	*Haas & Haynie Corp.*	GSBCA Nos. 5530, 6224, 6638, 6919-20, 84-2 BCA ¶ 17,446 (1984).	Office Building and Courthouse	Not specified		2, 3
B5	*Central Mechanical Construction*	ASBCA No. 29434, 86-3 BCA ¶ 19,240 (1986).	Fire Station Remodel	Not specified		2
B6	*Saudi Tarmac Co., Ltd and Tarmac Overseas, Ltd.*	ENGBCA No. 4841, 89-3 BCA ¶ 22,036 (1989).	Building Complex	Not specified		2
B7	*Freeman-Darling, Inc.*	GSBCA No. 7112, 89-2 BCA ¶ 21,882 (1989).	Modernize Cargo Inspection Facility	Delay Claim		3
B8	*McMillin Brothers Constructors, Inc.*	91-1 BCA ¶ 23,351, EBCA No. 328-10-84, 1990 WL 140900 (1990).	Research Facility (Civil/ Concrete)	"Novel modified total cost concept", where total labor overrun was claimed for 3 of 6 crafts		1, 2, 3, 4, 5
B9	*Atlantic Dry Dock Corp. v. United States*	773 F.Supp. 335, 338-339 (M.D.Fla., 1991).	Overhaul two Coast Guard ships	Not specified		2
B10	*Triple "A" South*	94-3 BCA ¶ 27,194, ASBCA No. 46,866 (1994).	Overhaul of a naval warship	Applied impact factors (variant of NAVSEA) associated with each major CO		2, 3, 5
B11	*Southwest Marine, Inc.*	94-3 BCA ¶ 27,102, DOTCAB No. 1663 (1994).	Repair of Coast Guard ship	Complex method of applying impact factors (variant of NAVSEA Guidelines); separate for local and cumulative disruption		3, 4, 5

No.	Case	Citation	Project/ (Discipline)	Damages Methodology	Notes on the Disruption Portion of Claim	Reasons for Rejection
B12	*Aetna Casualty & Surety Co. v. George Hyman Construction Co.*	U.S. Dist., LEXIS 22627 (E.D. Pa. 1998).	Rehabilitation of Railway Station	Modified Total Cost and Leonard Study	Subcontractor claim that magnitude and cumulation of number of COs amounted to a constructive change. Sub lost $0.7M on $3.3M base contract, of which $0.44 M was uncompensated labor costs. No substantial dispute that COs were principal cause of inefficiencies. Court not persuaded that magnitude of changes resulted in a separate impact, and also found that CO pricing included impact costs (accord and satisfaction).	1, 3, 4, 5
B13	*Centex Bateson Constr. Co.*	VABCA Nos. 4613, 5162, 5165, 99-1 BCA ¶ 30,153 (1998), aff'd, Centex Bateson Constr. Co. v. West, 250 F.3rd 761 (Fed Cir. 2000).	Medical Center (Electrical)	Quantified lost MHs by month and area using earned and actual productivities, no. and magnitude of events, schedule activities impacted, manpower, etc.	Claims based on 206 COs that added 12 percent to original contract value. Board acknowledged that it would grant recovery on a cumulative impact claim, but contractor failed to prove its claim. Board "...cannot find...that the number, scope, or timing of the changes was significantly at variance with what could be expected in a project of this magnitude."	1, 2, 3, 4, 5
B14	*Coates Industrial Piping, Inc.*	VABCA No. 5412, 99-2 BCA ¶ 30,479 (1999).	Mechanical Work at Medical Center	Delay Claim	Contractor claimed extended performance/impact costs due to the number and scope of changes after a given date (approx. 10 percent added to contract value). Claim rejected, as Board found no cumulative impact given number and character of changes, and that contractor improperly included impact costs in agreed COs.	3, 5
B15	*J. A. Jones Construction Co.*	00-2 BCA ¶ 31,000, ENGBCA Nos. 6390-1, 6386-7, 6348, 6388-9, 2000 WL 1014011.	Flood Wall and Ground Raising; Flood Protection	Measured Mile (Contractor abandoned its Leonard Study analysis)	On summary judgment, cumulative impact claim barred by release language in COs.	3, 4, 5
B16	*Bay Construction Co.*	VABCA Nos. 5594, 5625, 5626, 5628 and 5831, 2002 WL 442118 (2002).	Building	Referred to as "measured mile" (opposing expert argued it was not)	Appeals court ruled that contractor could not pursue cumulative impact claims associated with 2 large modifications (of 120 total) due to accord and satisfaction/ release language in agreed modifications.	3, 4, 5

No.	Case	Citation	Project/ (Discipline)	Damages Methodology	Notes on the Disruption Portion of Claim	Reasons for Rejection
B17	*Jackson Construction Co. v. United States*	62 Fed. Cl. 84 (Fed. Cl. 2004).	Building	Total Cost	Contractor sought delay and delay damages due to, in part, the cumulative impact of changes. 40 COs were issued, which increased the original contract by just under 10 percent. Board denied claim, not persuaded that magnitude and cumulation of changes significantly altered the contract.	2, 3, 5
B18	*Venetian Lien Litig.*	2004 WL 3265025 (Nev. Dist. Cr., 2004).	Building	Not specified	Board acknowledged that cumulative impacts do occur and in some instances should be compensated. However, claim was denied for the following reasons: changes were found to relatively minor in number and magnitude; delay/acceleration found to have been compensated through COs; alleged impacts were not supported by project record; no specific analysis of timing of impacts; and failure to establish quantum under total cost approach.	6
B19	*Strand Hunt Const., Inc. v. Lake Washington School Distr., No. 414*	Not Reported in PJd, 2006 WL 2536315 (Wash. App. Div. 1).	Building/ School	*Quantum Meruit*	Defendant granted summary judgment, which excluded plaintiff's further cumulative impact and other delay and disruption claims on agreed modifications due to release language.	2
B20	*Dugan & Meyers Constr. Co., Inc. v. Ohio Dept. of Adm. Servs.*	113 Ohio St.3d 226, 864 N.E.2d 68 (2007).	Building	Claimed total costs above adjusted bid for "general conditions" accounts.	Direct and cumulative impact claims associated with 600 formal COs that doubled the original contract price. Board rejected majority of quantum based on: (i) waived rights on agreed COs; (ii) lack of causal evidence; and (iii) unreliable quantum calculations. Minor award granted by jury verdict for local disruption costs.	1, 2, 6

No.	Case	Citation	Project/ (Discipline)	Damages Methodology	Notes on the Disruption Portion of Claim	Reasons for Rejection
B21	*Lake Union Drydock Company, Inc. v. U.S.*	United States District Court, W.D. Washington, at Seattle, No. C05-2146RSL (October 2007).	Ship Repair/ Conversion	NAVSEA Guidelines	Cumulative impact claimed associated with 202 change requests incorporated into 95 contract mods, which increased the original contract price by 64 percent (as claimed). Board denied claim in full: (i) largest mod was issued prior to start of work and other mods were relatively small; (ii) failure to prove causation; (iii) quantum calcs unreliable and failed to consider contractor bid error and performance problems; and (iv) properly calculated, the actual mh growth was 26 percent, which the Board found normal for similar overhaul contracts.	3
B22	*AMEC Civil, LLC v. DMJM Harris, Inc.*	Civil Action No. 06-64 (FLW), US District Court, D. New Jersey (June 2009).	Bridge	Actual Cost Overrun for Specific Cost Codes (Modified Total Cost)	Summary judgment wherein the court held that the contractor could not use the theory of cumulative impact or the modified total cost method to prove its damages. Rather, the contractor would have to demonstrate which of its specific damages were attributable to the alleged negligence.	1, 3, 4
B23	*Walsh/Davis Joint Venture*	CBCA No. 1460 (March 2012).	Building Complex	Not specified	Board ruled that contractor could not recover cumulative impact costs in respect of modifications which included certain language ("Settlement of this change includes all costs, direct, indirect, and impact and delay associated with this change"). Contractor could recovery for costs on modifications in which alternative language was used.	2

Appendix B – Court/Board Decisions not for the Contractor

No.	Case	Citation	Project/ (Discipline)	Damages Methodology	Notes on the Disruption Portion of Claim	Reasons for Rejection
B24	*Haskell Corporation, et al v. ConocoPhillips Company*	No. A124446, Court of Appeals of California, First District, Division Four. (March 2012).	Oil Refinery Revamp	Based on Change Requests that amounted to more than a Total Cost claim	Cumulative impact claim of $10.6 million for late design, thousands of RFIs, and late owner-provided foundations and equipment. Appeals court affirmed a trial court's decision that: (i) the contractor failed to prove abandonment of the contract; and (ii) cumulative impact claims were barred by language in a contemporaneous settlement agreement. Includes discussion of the contractor's failure to accurately track total costs and impact costs.	1, 2, 4, 5
B25	*Aragona Constr. Co. v. United States*	165 Ct. Cl. 382 (1964).	Hospital	Total Cost	65 formal COs and 37 other minor modifications did not amount to a cardinal change. Court found that changed work had been paid, and that changes related primarily to the substitution of materials.	1
B26	*Wunderlich Contracting v. U.S.*	173 Ct.Cl. 180, 351 F.2d 956 (1965).	Hospital	Total Cost	Contractor claimed total losses based on defective plans and specs. Agreed COs added 6 percent to base contract price; total overrun approximately 20 percent over base price. Board found changes not so substantial as to warrant breach/cardinal change. Contractor failed to differentiate and show effect of government-caused problems; total cost insufficient proof.	3, 5
B27	*Watt Plumbing, A.C., and Electrical v. Tulsa Rig, Reel, and Manufacturing Co.*	533 P.2d 980 (1975).	Hospital	*Quantum Meruit*	Contractor alleged breach of contract due to excessive changes/cardinal change. Court found that agreed modifications became part of the contract, and therefore cannot constitute breach of contract.	1, 2

No.	Case	Citation	Project/ (Discipline)	Damages Methodology	Notes on the Disruption Portion of Claim	Reasons for Rejection
B28	*John E. Green Plumbing & Heating Co., Inc. v. Turner Const. Co.*	742 F.2d 965, 968 (6th Cir. 1984).	Hospital (Mechanical/ Plumbing)	Arbitrary allocation of total overrun to 17 categories	Contractor claimed entire overrun of 98,000 mhs from base budgeted of 76,000 mhs, alleging out of sequence work, lack of temporary heating, and delayed approval of design changes. Court denied claims: Failure to prove damages; unable to separate CO work from other impacts; no basis for allocation of damages; therefore "proper to reject the entire claim."	4, 5
B29	*The Triax Co. v. United States*	28 Fed.Cl. 733 (1993), aff'd 19 F. 3d. 1196 (Fed. Cir. 1994).	Housing Renovation (486 units)	Total Cost	Number of items on changes forms totaled 22,175, yet were generally minor in scope and price. Plaintiff could not establish "causal connection" between changes and increased costs. Court found that plaintiff bid error and other factors may have led to cost growth. Term "cumulative impact" not used specifically.	3, 5
B30	*Hoffman Const. Co. of Oregon v. U.S.*	40 Fed.Cl. 184 (1998).	Building Renovation	Formula Used: [(Labor Overrun in Specific Areas) x (percent Attributed to Owner)]	A portion of claimed costs related to inefficiencies allegedly caused by multiple changes, late responses, and other delays. Plaintiff failed to prove causation and quantum, and executed a CO which released government from liability for portions of the work.	2, 3, 4, 5

No.	Case	Citation	Project/ (Discipline)	Damages Methodology	Notes on the Disruption Portion of Claim	Reasons for Rejection
B31	*Advanced Engineering & Planning Corporation*	ASBCA Nos. 53366, 54044 (November 2004.	Ship Alteration	Variant of NAVSEA Guidelines and a separate 20 percent impact factor	Disruption claimed for deficient specs, late/ incomplete information, formal and constructive changes, delay, and acceleration. Claim rejected for lack of causation, the board noting that the contractor provided virtually no testimony or documentation establishing causation. (Separately, the board awarded the contractor disruption costs on unresolved formal changes. The factors used in the change estimates were 5 percent for local disruption and 10 percent for cumulative disruption applied to extra mhs, to which the Navy did not object.)	3, 4

Appendix C

Engineering and Construction Project Documentation

(Hard Copy or Original/PDF & Native Electronic Files*)

Precontract Documents

- Addenda to ITB/Specifications
- Bid Clarification Letters/Responses
- Bidder "Question and Answer" Meeting Notes and Minutes
- Contractor's As-Bid Schedules and Manpower Curves*
- Contractor's Basis of Bid Estimate (take-offs, calculations, labor rates, etc.)*
- Contractor's Bid Estimate Worksheets*
- Contractor's Bid Labor Productivity Assumptions
- Contractor's Bid Submittal/Proposal (Technical and Commercial)
- Owner's Bid Evaluations
- Owner's Bid Package Drawings and Specifications
- Owner's Bid Schedule Showing Work by Other Contractors*
- Owner's Instructions to Contractor
- Owner's Invitation to Bid/Request for Proposals
- Owner's ITB Schedule*
- Pre-Bid Correspondence
- Pre-Bid Meeting Minutes & Notes
- Prime Contractor's Bid Evaluations of Subcontractor's Bids
- Prime Contractor's Bid Packages to Subcontractor's
- Project Labor Agreements
- Site Investigation Report/Notes
- Subcontractor Bids and Schedules*
- Vendor Quotations Used for Bid

Contract/Project Management Documents

- Change Order Procedure
- Change Orders and Contract Amendments (executed after the Contract initiation but then become part of the Contract)
- Contract Documents (General/Special Conditions, Exhibits)
- Construction Management Plan
- Drawings and Specifications Issued with Contract (if Different from ITB)
- Environmental Plan
- Notice to Proceed
- Procurement Plan
- Project Financing Plan
- Project Management/Execution Plan
- QA/QC Plan
- Risk Management Plan
- Safety Plan
- Scheduling Procedure
- Staffing Plan
- Start-up & Commissioning Plan
- Subcontract Documents (General/Special Conditions, Exhibits)
- Subcontracting Procedure
- Transportation Plan

Appendix C – Engineering/Construction Project Documentation

Contemporaneous Records

- Approved Vendor Lists
- Certificates of Occupancy
- Change Order Logs
- Construction Equipment Use Reports
- Construction Manager's Daily Reports
- Contractor's and Subcontractors' Weekly and Monthly Progress Reports
- Conversation Memos
- Customs Reports
- Daily Reports by Superintendents
- Diaries
- Drawing Revisions/Logs
- Electrical Drawings
- Equipment Rental Records
- Field Sketches
- File Indices
- Final Completion Certificates
- General Arrangement Drawings
- Hydrotest Reports
- Inspection Reports
- Installed Equipment Lists
- Material Receiving Reports*
- Material Receiving Tickets
- Mechanical Completion Certificates
- Nonconformance Reports
- P&IDs
- Pipe Spool Fabrications and Delivery Reports/Logs
- Piping Isometric Drawings
- Plot Plans
- Prime/General Contractor's Daily Construction Reports
- Process Flow Diagrams
- Progress and other Meeting Minutes
- Progress Photographs, particularly of problem activities, including date, description of photograph, and name of photographer
- Punchlists
- Requests for Information & Logs and Owner's/Engineer's Responses*
- Shop Drawing Logs
- Specification Revisions/Logs
- Structural Drawings
- Subcontractor Daily Reports
- Submittal Logs
- Substantial Completion Certificates
- Systems Completion/Turnover Certificates
- Telephone Logs
- Test Reports/Certificates
- Transmittal Logs
- Weather Data

Schedule Information

- Bar Chart Schedules*
- Critical-Path Schedules (Original and Updates)*
- Manpower Loading Charts*
- Scheduling Meeting Minutes
- Short-Interval Detailed Schedules*
- Time Extension Request, Impact & Delay Schedules*

Correspondence

- Conversation/Telephone Memorandums
- Correspondence Logs
- Emails between Contractor/Subcontractor
- Emails between Owner/Contractor
- Emails to/from Construction Manager
- Emails to/from Owner's Project Management Representative
- Emails with Design Firm/Architect
- Emails with Vendors
- Handwritten Notes
- Internal Office Memos and Correspondence
- Interoffice emails in Contractor's/Subcontractor's Organizations
- Interoffice emails in Owner's Organization

Correspondence (continued)

- Letters and Other Communications to/from Vendors
- Letters to/from Contractor
- Letters to/from Design Firm/Architect
- Letters to/from Owner
- Letters to/from Owner's Project Management Representative
- Letters to/from Subcontractors
- Notices of Changes or Problems to Contractor from Subcontractors
- Notices of Changes or Problems to Owner and to Owner's Project Representative
- Progress Meeting Minutes

Cost/Man-hour/Quantity Data

- Bank Loan Documentation
- Basis of Liquidated Damages Amounts
- Bulk Material Quantity Reports (Planned and Actual)*
- Canceled Accounts Payable Checks
- Canceled Payroll Checks
- Cash Flow Projections
- Certified Payroll Reports*
- Change Orders with Calculations, Estimates, & Cost Documentation (Approved, Rejected, Pending)
- Code of Accounts (Contractor & Subcontractor)
- Contractor's Home Office Overhead Reports (Revenue & G&A costs over the life of the contract)
- Currency Exchange Rate Data
- Daily Equipment Cost Records, including hours operated, hours idle, work performed, and repairs
- Labor Timecards Coded to Work Activities
- Labor Wage and Fringe Agreements
- Overtime Labor Records
- Paid Invoices
- Prime Contractor's Control Budget*
- Prime Contractor's Job Cost and Manpower Reports (Initial, Weekly, Monthly, Final)*
- Prime Contractor's Weekly Labor Distribution Reports (Initial, Weekly, Monthly, Final)*
- Productivity/Earned Value Reports
- Progress Payment Requests/Reports
- Purchase Orders
- Revenue Records
- Subcontractor Job Cost and Man-hour Reports (Initial, Weekly, Monthly, Final)*
- Subcontractor's Control Budgets*

Arbitration/Litigation Documents

- Arbitration/Court Orders & Timetables
- Claim Documents
- Counterclaim Documents
- Deposition Transcripts
- Document Production Requests
- Expert Reports and Rebuttal Expert Reports
- Interrogatories
- Pleadings, Briefs, Statement of Claim, Statement of Counterclaim, Statement of Defense, and all Replies and Responses
- Responses to Document Production Requests
- Responses to Interrogatories
- Witness Statements

Index

Case Citations

Cases

www.ingramcontent.com/pod-product-compliance
Lightning Source LLC
LaVergne TN
LVHW060727210926
848485LV00006B/1398